SpringerBriefs in Applied Sciences and Technology

Series editor

Janusz Kacprzyk, Polish Academy of Sciences, Systems Research Institute, Warsaw, Poland

SpringerBriefs present concise summaries of cutting-edge research and practical applications across a wide spectrum of fields. Featuring compact volumes of 50–125 pages, the series covers a range of content from professional to academic.

Typical publications can be:

- A timely report of state-of-the art methods
- An introduction to or a manual for the application of mathematical or computer techniques
- A bridge between new research results, as published in journal articles
- A snapshot of a hot or emerging topic
- An in-depth case study
- A presentation of core concepts that students must understand in order to make independent contributions

SpringerBriefs are characterized by fast, global electronic dissemination, standard publishing contracts, standardized manuscript preparation and formatting guidelines, and expedited production schedules.

On the one hand, **SpringerBriefs in Applied Sciences and Technology** are devoted to the publication of fundamentals and applications within the different classical engineering disciplines as well as in interdisciplinary fields that recently emerged between these areas. On the other hand, as the boundary separating fundamental research and applied technology is more and more dissolving, this series is particularly open to trans-disciplinary topics between fundamental science and engineering.

Indexed by EI-Compendex and Springerlink.

More information about this series at http://www.springer.com/series/8884

Yuri N. Toulouevski · Ilyaz Y. Zinurov

Fuel Arc Furnace (FAF) for Effective Scrap Melting

From EAF to FAF

 Springer

Yuri N. Toulouevski
Holland Landing, ON
Canada

Ilyaz Y. Zinurov
Akont
Gipromez
Chelyabinsk
Russia

ISSN 2191-530X   ISSN 2191-5318   (electronic)
SpringerBriefs in Applied Sciences and Technology
ISBN 978-981-10-5884-4   ISBN 978-981-10-5885-1   (eBook)
DOI 10.1007/978-981-10-5885-1

Library of Congress Control Number: 2017948619

Printed on acid-free paper

This Springer imprint is published by Springer Nature
The registered company is Springer Nature Singapore Pte Ltd.
The registered company address is: 152 Beach Road, #21-01/04 Gateway East, Singapore 189721, Singapore

Introduction

The purpose of writing this small book is to justify the need to create a new type of steelmaking unit namely Fuel Arc Furnace (FAF). The main feature of the FAF is high-temperature scrap preheating by powerful oxy-gas burner devices in combination with melting a scrap in the liquid metal. Implementation of the FAF is the promising direction of further development of EAFs. It is this direction that is capable of providing the deepest replacement of electrical energy by the energy of fuel and a further increase in productivity at maximum efficiency.

At present, there are very favorable conditions for the creation of the FAF. Thanks to the new methods of producing shale gas, its price has fallen sharply and the possibility of using increased significantly. In addition, development and implementation of shaft furnaces of the Quantum and COSS type contributes to the implementation of the FAF. This significantly facilitates the FAF development since the shaft is an optimal device for scrap preheating, and a new furnace can be created on the advanced design basis which has already implemented. The combination of high-temperature scrap preheating with the process of scrap melting in the metal bath can be able to provide (without an increase in furnace capacity and transformer power) much higher productivity and almost twice reduced electrical energy consumption in comparison with modern EAFs. This fact is confirmed by the calculations based on the reliable experimental data and simplified physical process models. These calculations are an important feature of the book. They greatly contributed to a better understanding of the main dependences between the key parameters of shaft furnaces.

The authors hope that their new book will encourage an interest in the problems of the creation of the FAF and the concentration of efforts in this direction.

The book may be useful not only to developers of new technologies and equipment for EAFs but also other specialists-metallurgists and students studying metallurgical specialties.

The problems of developing FAF were discussed by the authors with many specialists at the plants, the companies, and the design bureaus which contributed to a better understanding of this problem. The authors express their deep gratitude to all of them. The authors thank Dr. Christoph Baumann for his constant attention and support of this work. Special thanks go to Galina Toulouevskaya for her extensive work on preparation of the book for publication.

Contents

Chapter 1
EAF in Global Steel Production; Energy and Productivity Problems

Abstract In modern EAFs electrical energy consumption is on the average about 375 kWh per ton of liquid steel. Such high electrical energy consumptions cannot be justified as EAFs have great potentials of a deep substitution of much cheaper and affordable energy of natural gas for electrical energy. The entire heat in an EAF can be divided into three stages: the heating of scrap up to an average mass temperature of 1000–1100°; further heating and melting down of scrap; and finally, heating the melt to a tapping temperature. At the typical tapping temperature, the enthalpy of liquid metal amounts to on the average 400 kWh/t; and the enthalpy of scrap at a temperature of 1050 °C is about 200 kWh/t. Thus, 50% of the total energy transferred to the scrap and the liquid bath during the process is consumed at the first stage where, unlike the subsequent stages, the use of electrical energy is not absolutely necessary. To heat a scrap to a temperature of close 1000 °C it is necessary to use the energy of fuel instead of electrical energy. This would allow reducing electrical energy consumption by a factor of 1.8.

Keywords Electrical energy consumption in EAFs · Replacement with fuel energy · Enthalpy of scrap and liquid metal

1.1 Production of Steel from Scrap Is EAF's Mission

There are two basic ways of steel production which are characterized by different sources of iron. In the first way iron ore is used as a raw material, in the second one, metal scrap is recycled. Iron ore either is converted into liquid iron (hot metal) by using energy of coke in a blast furnace or other way it is subjected to direct reduction with obtaining DRI and HBI in the form of briquettes or pellets. In oxygen convertors the hot metal is converted to steel by means of oxygen blowing and by using chemical and physical energy of the hot metal itself. Scrap as well as products of direct reduction are melted down and converted into liquid steel in electric arc furnaces (EAFs). In these furnaces, electrical energy is used as the main source of energy.

© The Author(s) 2017
Y.N. Toulouevski and I.Y. Zinurov, *Fuel Arc Furnace (FAF) for Effective Scrap Melting*, SpringerBriefs in Applied Sciences and Technology, DOI 10.1007/978-981-10-5885-1_1

In 2015, the total crude steel production in all countries over the world except China, with its very special terms, was about 806 million tons. 50% of this amount of steel has been produced in EAFs. In the US, the share of electric steel was 63%, 65% in Turkey, 70% in Mexico, and 78% in Italy [1].

Such a development of EAFs is determined by the presence of large stocks of scrap in advanced countries. These stocks are constantly renewed. Scrap is an optimal raw material for EAFs. The best performances of these furnaces are achieved when using 100% scrap in a charge. The allowable content of scrap in the charge of oxygen converters does not exceed on the average of 25%. Therefore, an oxygen converter is not a competitor to an EAF in terms of a scrap consumer. At the same, time more dense and expensive scrap is used in oxygen converters, and a relatively cheap scrap of inferior quality is processed in EAFs.

The use of scrap to produce steel has important environmental advantages. In integrated plants where steel is made of hot metal in oxygen converters the specific CO_2 emissions per ton of steel exceed those from EAFs in mini-mills operating on scrap by more than three times. This is explained by the fact that integrated plants comprise of blast furnaces as well as workshops producing agglomerate and coke which pollute the atmosphere by emissions not only of CO_2 but also of toxic gases such as CO and SO_2.

A significant disadvantage of scrap hindering the production of some high-quality steel grades is the contamination of scrap with copper, nickel, chromium and some other impurities. These impurities cannot be removed during the processes for production of finished steel. The allowable content of impurities is sharply limited for some steel grades. This obstacle is eliminated by a more thorough preparation of scrap for melting and also by the substitution, in part, of hot metal or direct reduction products for scrap.

Not only reduction in the concentration of harmful impurities but also a significant decrease in electrical energy consumption and tap-to-tap times are achieved by the substitution, in a part, of hot metal for the scrap. There are examples of the use of such a technology in some regions with the availability of hot metal and the acute shortage of scrap. However, this technology is unpromising because EAFs do not adapt to work with hot metal, and when using it they lose their advantages over oxygen converters. With the increase in a share of DRI and HBI in the EAF charge electrical energy consumption and tap-to-tap times increase significantly. Therefore, neither hot metal nor direct iron reduction products[1] can be considered as an alternative to scrap. Production of steel from scrap in EAFs accords with the common long-term strategy of the development of industry. As per this strategy the demand on source materials must be met most of all by returning to the production of materials obtained by recycling depreciated industrial products and waste.

---

[1]The global production of these products was 75 million tons in 2013.

1.2 Melting a Scrap as a Key Process of the Heat

Already for a long time, in EAFs as well as in oxygen converters a semi product with given temperature and carbon content is usually melted. This metal is reduced to a final chemical composition, cleaned of dissolved gases and nonmetallic inclusions, and heated up to an optimal temperature for casting conditions by means of ladle-furnaces, vacuum degassers, and other units of secondary metallurgy. Under these conditions, the scrap melting process is the main process of the heat which determines basic performances of the furnace. This is confirmed by energy consumption in the process and its relative duration.

Enthalpy (heat content) of semi-product at the tapping temperature amounts to on the average 400 kWh/t. In the course of the heat the melt and scrap obtain this amount of heat from all sources of energy such as: electric arcs; burners; carbon monoxide (CO) released from the bath; and chemical exothermic reactions in a liquid bath. About 92% of all the heat obtained is consumed for heating and melting down of the scrap and 8% only for heating the melt from the melting point to the tapping temperature, Table 1.1. The melting period time in modern EAFs amounts to more than 80% of the power-on furnace operation time.

1.3 Unjustified High Electrical Energy Consumption

In modern EAFs electrical energy consumption is on the average about 375 kWh per ton of liquid steel. Costs for electrical energy and electrodes[2] are very close to the cost of metallic charge occupying the first place in the list of costs. This significantly increases the cost of products of mini-mills and reduces their competitiveness.

High electrical energy consumptions cannot be justified as EAFs have great potentials of a deep replacement of electrical energy with much cheaper and affordable energy of natural gas. A long expected sharp drop in gas prices creates particularly favorable conditions for such a substitution. The electrical energy has to be used for solving only those problems which cannot be resolved by using fuel energy. In this connection, let us refer to Table 1.1.

The entire heat in an EAF can be divided into three stages: the heating of scrap up to an average mass temperature of 1000–1100°C; further heating and melting down; and finally, heating the melt to a tapping temperature. At the typical tapping temperature, the enthalpy of liquid metal amounts to on the average 400 kWh/t; and the enthalpy of scrap at a temperature of 1050 °C is about 200 kWh/t. Thus, 50% of the total energy transferred to the scrap and the liquid bath during the process is consumed at the first stage where, unlike the subsequent stages, the use of electrical energy is not absolutely necessary.

---

[2]Electrode consumption is increasing along with electricity consumption.

Table 1.1 Heat capacity c and enthalpy E of solid and liquid iron at various temperatures t

t, °C	c, Wh/kg °C	E, kWh/t	t, °C	c, Wh/kg °C	E, kWh/t
25	0.124	3.1	900	0.190	171
50		6.5	950		181
100	0.130	13.0	1000	0.190	190
150		19.8	1050		198
200	0.134	26.8	1100	0.188	207
250		34.1	1150		216
300	0.139	41.6	1200	0.187	225
350		49.5	1250		234
400	0.144	57.7	1300	0.187	243
450		66.5	1350		254
500	0.155	77.5	1400	0.190	266
550		85.0	1450		275
600	0.158	95.1	1500	0.191	287
650		106	1536 solid		294
700	0.169	119	1536 liquid	0.240	369
750		132	1600		384
800	0.181	145	1640	0.240*	394
850		158	1680		403

Notes Enthalpy of low-carbon scrap is higher than those by 2–3%
*Average heat capacity over a range of temperatures 1536–1680 °C

Over the world, the heating of steel billets and products up to a temperature of 1200–1300 °C for subsequent rolling, forging and thermal processing to make is carried out, with a few exceptions, in furnaces heated by natural gas. In EAFs, the fuel energy has to be used as well when heating a scrap up to a temperature close to 1000 °C. This would allow reducing electrical energy consumption by about 1.8 times.

1.4   Problems of Ultra-High Power (UHP) EAFs with Regard to Energy

Let us consider problems related to high electrical power of EAFs. In the past, it was the increase in the electrical power of EAFs that resulted in dramatic shortening tap-to-tap times and an increase in productivity of furnaces to the current level. Without such anincrease in productivity the EAF could have never become the very steelmaking unit which along with oxygen converter is a determinant of world steelmaking. The highest productivity is achieved by the concurrent increase in electrical power and capacity of the EAF.

Electric arc furnaces are mostly intended to be installed at mini-mills where they determine productivity of the entire plant. It is most reasonable to equip steel-making shops at such plants with one furnace. Such organization of production allows minimizing manpower and operating costs in general. In addition, in the shops equipped with a number of furnaces any disruption of the preset production pace at one of the furnaces causes delays of other furnaces as well thus reducing their overall productivity. For these reasons, in those cases when it is required to provide an annual mini-mill output in excess of 2 million tons one 300–400-t furnace is preferred to be installed instead of several EAFs of smaller capacity. Performances of these furnaces can be seen in the following examples.

At the mini-mill in Gebze (Turkey) a furnace of 420 t capacity and tapping weight of 320 t operates instead of four 55-t EAFs. The transformer power is 240 MVA with an option to increase by 20%. During the melting period the maximal active power of arcs reaches 200 MW. Productivity exceeds 320 t/h. Electrical energy consumption is 362 kWh/t [2]. At the mini-mill in Iskendederun (Turkey) an EAF of 300 t capacity with tapping weight of 250 t and transformer power of 300 MVA is installed. The productivity is 320 t/h, 2.4 million tons per year [3].

This extensive development of EAFs is not rational. Such UHP EAFs require the use of the most expensive electrodes 710 and 820 mm in diameter and, along with other problems, make very high demands on the power supply system and to the electrical grids. This creates considerable difficulties in the construction of mini-mills in regions where the electrical network does not meet such high demands. For example, in order to feed a 420-t EAF of 175 MW power operating near Tokyo, Japan, a direct current had to be used despite the rise in the cost of energy source. This is explained by insufficient capacity of power grids in this area. An AC furnace of the same power would create too large electrical disturbances in these grids to other consumers of electricity [4].

The required in each case a higher level of productivity can be achieved not only due to corresponding increase in capacity and in electrical power of the furnace but also by intensifying the scrap melting process using a more efficient means such as a high-temperature scrap preheating and intensive bath stirring. The intensive direction of development is always much more effective than extensive one. This general rule holds true for EAFs as well. There is no doubt, that any mini-mill at the same productivity will prefer to use a furnace of smaller capacity and lower electrical power.

1.5 High Productivity or Low Costs?

Recently, the articles devoted to the shaft furnaces have been discussing the issue: what is the main goal - high productivity or low costs?

Such a contraposition of productivity and costs seems to be unjustified and inconsistent. It does not contribute to the choice of optimal direction of further development of EAFs. This direction cannot be related to downswing or upswing in production, which periodically alternate. At one and the same time, in some countries there may be a decline, and in others a significant rise. The combination of an increase in productivity with reduction in costs is a sign of the optimal direction of development. This is confirmed by the entire history of the EAF development. Almost all of the innovations implemented in EAFs for the last 30–50 years were directly or indirectly aimed to an increase in productivity, but at the same time they have resulted in a sharp reduction in the cost of electricity, electrodes and refractory.

An increase in productivity is one of the most efficient means of reducing the production costs. At relatively low productivity, the new types of furnaces will not be able to successfully compete with the best modern EAFs. Furnaces of the future should have such a set of main performances which would force all other types of EAFs to abandon. This means both a much higher productivity and much lower costs for steelmaking from scrap.

References

1. Steel Statistical Yearbook (2016) worldsteel Committee on Economic Studies, Brussel p. 46. http://www.worldsteel.org
2. Abel M, Hein M (2008) The Simetal Ultimate at Colakoglu, Turkey. In AISTech 2008 Conference. Pittsburgh, Pa., USA
3. Sellan R, Fabbro M, et al (2008) The 300 t EAF meltshop at the Iskenderun mini-mill complex. MPT Internation (2): 52–58
4. Mukkhopadhyay A, Coughlan R, et al (2012) An advanced EAF optimization suite for 420-t jumbo DC furnace at Tokyo steel using DANIELI technology. In AISTech Conference, Proceedings, vol 1. pp 745–756

Chapter 2
Analysis of Technologies and Designs of the EAF as an Aggregate for Heating and Melting of Scrap

Abstract The following issues are considered: the role of hot heel in scrap melting by electric arcs in the furnace freeboard; advantages and disadvantages of furnaces with a single charging and those with a telescoping shell; Specifics of furnace scrap hampering its heating by burners. The possibilities of using different types of burners for scrap heating are analyzed including stationary burners and jet modules as well as slag door, oriel, and roof rotary burners. The data of industrial tests of the process of two-stage melting of scrap in different EAFs with the use of rotary burners without using electrical power in the first stage are given. Under conditions of short tap-to-tap times in modern EAFs the high-temperature scrap heating by burners is impossible. The advantages and disadvantages of EAFs of various types with preheating of scrap with off-gases and melting of heated scrap in liquid metal, including both Consteel conveyer furnaces and shaft furnaces of the Quantum, SHARC, COSS, ECOARC types are considered. In all the shaft furnaces, the scrap preheating temperatures do not exceed 400–450 °C and electrical energy consumption is about the same equal to 300 ± 15 kWh/t. This is explained by the fact that the possibilities of further raising the scrap preheating temperature with off-gases and thereby reducing the consumption of electrical energy are practically exhausted.

Keywords Stationary and rotary burners · EAF types with scrap preheating · Two-stage process

2.1 Melting a Scrap by Electric Arcs. Function of Hot Heel

In a conventional technology, scrap is charged by baskets from the top and placed in a furnace freeboard where it is mainly melted by electric arcs with little involvement of burners and other energy sources. Direct contact of scrap pieces with arcs plasma having a temperature close to 6000 °C provides a high melting rate which increases with increasing power of arcs.

© The Author(s) 2017
Y.N. Toulouevski and I.Y. Zinurov, *Fuel Arc Furnace (FAF) for Effective Scrap Melting*, SpringerBriefs in Applied Sciences and Technology, DOI 10.1007/978-981-10-5885-1_2

With the increase in EAF's electrical power, the process of the heat with hot heel where a quantity of metal and slag at each tapping is left on the bottom received general use. In high power furnaces, boring-in scrap pile occurs so quickly that the layer of melt is not deep enough when electrodes reach closely to the bottom. Before, in the absence of hot heel, there was a danger of damaging bottom refractory by powerful arcs. This factor restricts increasing electrical power of the furnaces. The hot heel has eliminated the said limitation and allowed increasing electrical power with the aim of further increase in productivity.

In furnaces with hot heel, scrap discharged from the bottom of baskets is immediately immersed into a bath and melted in liquid metal. With increasing the hot heel weight a share of the charge melted in this manner increases. This fact has to be taken into consideration in calculations of melting time of the entire scrap. Operation with the hot heel has also a number of technological advantages such as tapping without slag, early start of bath blowing with oxygen, submerging arcs into a foamed slag, etc. To maintain the mass of metal retained in the furnace at a relatively constant level close to the optimum is a necessary condition for complete and stable enough to use all the advantages of operation with hot heel. For this purpose, furnaces are equipped by sensors which allow controlling a furnace weight varying during the heat and consequently a hot heel weight.

2.1.1 Single Scrap Charging

Recently, EAFs with expanded freeboard size capable of receiving all scrap of about 0.7 t/m^3 bulk density charged by single basket are getting spread. Charging each basket requires roof swinging and current switching off. With short tap-to-tap time, using one scrap basket instead of two leads to a considerable increase in EAF's hourly production. However, the advantages of furnaces with single scrap charging are not limited to that.

Freeboard volume is expanded in such furnaces mainly by means of increasing its height. Greater height of scrap pile in the furnace provides for better scrap absorbing the heat of hot gases, obtained when post-combusting of CO, passing upwards through scrap layer from below. The same can be said about absorbing heat from flames of oxy-gas burners installed in the lower parts of furnace sidewalls. Increasing depth of pits bored-in by arcs in scrap also increases the degree of arc heat assimilation. All this increases scrap heating temperature prior to its immersion into the melt, and accelerates melting. At the same time, electric energy consumption is decreased due to the reduction in the time when the furnace is open and loses a lot of heat. Dust-gas emission into shop atmosphere is reduced also while scrap charging. In furnaces of 300–400 t capacity, due to the freeboard height extend, the number of charges is decreased to two per heat.

However, considering the effect of increasing the furnace freeboard height on the utilization of heat in it, it should be taken into account that sidewall area is increased and, consequently, heat losses with cooling water are increased as well. To reduce these losses, measures are taken to increase the thickness of skull layer on the sidewall panels. For instance, Danieli Company uses panels consisting of two layers of pipes. The pipes of the internal (with respect to freeboard) layer are spaced apart much wider than in the external one. That facilitates formation of thicker skull layer and its better retention on the pipes. As freeboard height is increased considerably, electrode stroke and their length are respectively increased as well, thus increasing the probability of electrode breaking. To prevent breaking the rigidity of arms and of the entire electrode motion system should be increased. The lateral surface area of electrodes as well as their wear due to oxidation, which is about 50% of the total electrode consumption, is increased. Taking into consideration all these factors it can be assumed that furnaces of no more than 180 t capacity are most suitable to realize a single scrap charging.

It should be paid attention to the fact that a freeboard height required during the scrap charging and the initial melting stage comes into conflict with an optimum height after the flat bath formation when this height should be significantly short-ened in order to reduce heat losses. In EAFs with a single charging this contra-diction is considerably enhanced.

2.1.2 Telescoping Shell

In order to eliminate the above drawback of single charging furnaces it is necessary to periodically during the heat reduce the freeboard height down to a minimum in the melting end. An EAF with a variable freeboard height has been developed by the Company Fuchs Technology AG. In this furnace, the height reduction is achieved by means of lowering of the roof. The main advantage of this design is that when single charging it allows using a lower-cost scrap with lower bulk density at lower heat losses with water. Operation of such a furnace in a mini-mill in Turkey has shown the ability to reduce a scrap density to 0.55–0.60 t/m^3 while reducing electrical energy consumption by about 2%. It should be noted that in furnaces with scrap charging by separate portions into a liquid bath, Sect. 2.2, there is no necessity to vary the freeboard height in the course of the process.

2.2 Heating a Scrap by Burners in the Furnace Freeboard

2.2.1 Specifics of Furnace Scrap Hampering Its Heating

In EAF, as a rule, the cheapest light scrap is used. It usually has a low bulk density of 0.6–0.7 t/m^3. Such a scrap consists mostly of lumps with relatively small mass and thickness. The length and shape of these lumps vary widely. The denser, cleaner and more expensive scrap is used in converters which are not suitable for melting light scrap. Intent of metallurgists to use cheap scrap in EAF is determined by the fact that cost of scrap accounts for approximately 70% of total cost per heat of materials, energy and personnel.

Depending on the source of scrap supply and the method of its preparation for melting the thickness of scrap lumps varies from a few millimeters (sheet bushelling) to 100–120 mm. Internal thermal resistance of such lumps is so low that each single lump can be heated at any practically achievable rate. The temperature difference between the surface of a lump and its centre remains negligible and can be ignored. This is not true for ladle skulls, trimmings of large ingots and other similar materials which are heated through quite slowly, and therefore their use should be avoided.

Though the scrap for EAF is preselected, it always contains some amounts of rubber, plastics and other flammable organic materials including oil. The chips from metal cutting machines are especially contaminated with oil. Oil and other flammable contaminants present in the scrap emit a lot of heat while burning out. This causes quite undesirable consequences. Even when moderate-temperature (1300–1400 °C) flame and gas is used for pre-heating of scrap, pockets of burning and melting of small fractions can be formed in the heated layer. When this occurs, the separate scrap lumps can be welded together forming so called "bridges" which obstruct the normal course of the melting process.

In the temperature range 400–600 °C oil and other organic materials contained in the scrap sublimate and burn releasing badly smelling toxic gases so-called dioxins, which requires serious measures of protection of the atmosphere of a shop and as well as environmental protection. At temperatures higher than 800–900 °C the fine scrap is oxidized intensely due to its very large surface area. This decreases the yield. The interaction of combustion products with a highly heated scrap is accompanied by their reduction and fuel underburning. Thus, the specifics of the furnace scrap utilized in EAF create certain difficulties for its heating, especially for the high-temperature heating.

2.2.2 Stationary Burners and Jet Modules

Low-power oxy-gas burners are widespread in EAFs. Unit power of such burners does not exceed 3.5–4.5 MW. They are installed in the wall panels, usually about 500 mm above the bath sill level, as well as in the oriel covers and in the slag doors. In the past, three sidewall burners used to be installed in the furnace in the so-called cold zones between the electrodes where the scrap melting required longer time. The sidewall burners equalized the temperature field along the whole perimeter of the furnace. The oriel burners eliminate the cold zone at the oriel, and the door burners do the same at the slag door sill area. The latter makes possible an earlier metal sampling and temperature taking, which allows shortening a heat. As burners had low unit power their use did not significantly affect electrical energy consumption.

Further practice has lead to understanding the necessity of increasing the fuel consumption in the burners not so much for the purpose of saving electrical energy as for intensification of the process. With tap-to-tap time being continuously reduced, this required a significant increase in the power of the burners. However, all attempts made in this direction have not given positive results. At present, unit power of burners, due to the reasons discussed in detail below, remains at the same level as 30–40 years ago. Therefore, in order to increase overall power of the burners, the number of burners has been increased. The number of burners in the furnaces reached six to nine, and in some cases even to 12.

Despite the increase in the number of burners, specific consumption of natural gas in the furnaces did not grow significantly. Usually, it does not exceed 8–10 m^3/t. This is a result of the further reduction of the tap-to-tap time and, correspondingly, burners' operation time. The effectiveness of the burners did not change as well. As before, they ensure reduction of tap-to-tap time and electrical energy consumption by not more than 6–8%.

The majority of burners under consideration despite a furnace size and their location are similar in general principles. Their design provides for intense mixing of gas and oxygen partially inside the burner and mostly close to its orifice. When used for scrap heating, the burners operate with oxygen excess coefficient of approximately 1.05. Usually, they form a narrow high-temperature flame. Initial flame speeds are close to the sonic speed or exceed it; maximum flame temperatures reach 2700–2800 °C.

Heating of liquid bath with burners is ineffective. However, small amounts of both gas and oxygen have to be supplied to the burners to maintain the so-called pilot flame. This allows to avoid clogging of the burner nozzles with splashed metal and slag. These forced non-productive consumptions of gas and oxygen noticeably worsen burners' performance indices.

Let us review the causes hindering the increase in power of stationary burners. During the operation of these burners, the direction of flame remains constant. Burner flames attack the scrap pile from the side, in the direction close to radial. The kinetic energy of the flames is low due to their low power. Penetrating into a layer of scrap these flames quickly lose their speed and are damped out. Therefore, their action zones are quite limited.

Since emissivity of oxy-fuel flame in the gaps between the scrap lumps is low, heat from flames to scrap is transferred almost completely by convection. With convection heat transfer, the amount of heat transferred to scrap per unit time is determined by: the surface area of the scrap lumps surrounded by gas flow; the speed of gas flow which determines the heat-transfer coefficient; and the average temperature difference between gases and heat-absorbing surface of the scrap. In the action zone of the burners, at high temperatures of oxy-gas flame the light scrap is heated very quickly to the temperatures close to its melting point. Then the scrap settles down and leaves the action zone of the flame which loses the convective contact with the scrap. In the course of the burners operation, the area of the heat-absorbing surface of the scrap lumps and the temperature difference between the scrap and the flame diminish progressively. The heat transfer remains high only during a short period after the start of the burners operation. Then the heat transfer reduces gradually and finally, drops so low that the burners must be turned off, as their operation becomes ineffective.

Besides, potential duration of burners operation is also limited by the physical-chemical factors. At the scrap temperatures approaching 1450 °C and especially during the surface melting of scrap, the rate of oxidation of iron by the products of complete combustion of fuel sharply rises. In doing so, the products of fuel combustion are reduced to CO and H_2 according to the following reactions:

$$CO_2 + Fe = FeO + CO \quad \text{and} \quad H_2O + Fe = FeO + H_2$$

The fuel underburning increases, and CO and H_2 burn down in the gas evacuation system. The temperature of the off-gases rises sharply which, along with the other signs of reduced effectiveness of the burners operation, requires turning the burners off.

The described above processes in the scrap pile attacked by a narrow high-temperature flame explain comprehensively the futility of attempts to increase the power of considered burners. In accordance with well-known aerodynamic principles, the length and the volume of the flame and, therefore, its action zone increases insignificantly as the power of oxy-gas burner increases. As a result, the critical temperatures causing fuel underburning and settlement of the scrap in this zone are reached in a shorter time. Respectively, approximately proportionally to the increase in power of the burner, the potential effective burner operation time is shortened, whereas the amount of heat transferred to the scrap increases insignificantly. Only a relatively small portion of scrap pile is heated, which has little effect on energy characteristics of the furnace.

In addition to burners the tuyeres for oxygen bath blowing and injectors for carbon powder injection into the bath to form a foamed slag and reduce FeO were also installed in sidewall panels of EAFs. As a result of the improvement of these systems, they were combined in multifunctional devices, the so-called jet modules. All structural elements of the modules are usually placed in water-cooled boxes protecting these elements from high temperatures as well as from damage during scrap charging. The boxes are inserted into the furnace through the openings in the sidewall panels, which considerably decreases the distances from the nozzles of the burners and from injectors to the bath surface. There is a wide variety of design versions of the jet modules. The advent and development of this direction is associated with the PTI Company (USA) and with the name of V. Shver.

Let us examine the arrangement of the module by the example of a typical design of PTI. Due to a higher durability this module compared to other modules can be installed closer to the sill level. Thus, a distance from the oxygen burner nozzle to metal surface does not exceed of 700 mm. Reducing oxygen jet length improves oxygen efficiency. This is a substantial advantage of the PTI module. Further, this design and the similar ones have gained wide acceptance in many countries around the world.

The PTI module contains the water-cooled copper block (1) in which the oxy-gas burner (2) with water-cooled combustion chamber (3) and the pipe (4) for the injection of carbon powder are located, Fig. 2.1. The burner (2) has two operating modes. In the first mode, it is used as a burner for heating of scrap and operates at its maximum power of 3.5–4.0 MW. The gas mixes with oxygen and partially burns within the combustion chamber (3). At the exit from the chamber, the high-temperature flame is formed, which heats and settles down intensively the scrap in front of the burner. The combustion chamber protects, to a considerable extent, the burner nozzles from the clogging by splashes of metal and slag.

In the second operating mode, the burner is mainly used as a device for blowing of the bath. The gas flow rate considerably decreases, and the oxygen flow rate sharply increases. In this case a long-range supersonic oxygen jet is formed. In this mode, the function of the burner alters. It is reduced to the maintenance of the low-power pilot flame. This flame shrouds the oxygen jet increasing its long range, prevents flowing of the foamed slag into the combustion chamber, and protects the nozzle of the burner from clogging as well.

The burner is controlled by a computer which switches its operating modes in accordance with the preset program. Immediately after scrap charging, the first

Fig. 2.1 Jet module (designations are given in the text)

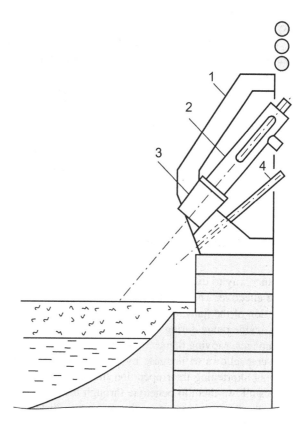

mode is switched on. In several minutes, it is switched to the second mode. The highly heated scrap can be cut by oxygen considerably easier than cold scrap. Therefore, the preliminary operation of the burner in the first mode greatly facilitates penetration of the supersonic oxygen jet through the layer of scrap to the hot heel on the bottom. This ensures the early initiation of the blowing of the liquid metal with oxygen, which is the necessary condition for achievement of high productivity of the furnaces. While the upper layers of scrap continue to descend to the level of the burner, the alternation of the operating modes is carried on and is repeated after charging of the next portion of scrap. This considerably increases the effectiveness of the use of oxygen in the initial period of the heat before the formation of the flat bath.

The module operating reliability in a decisive measure depends on durability of the protective boxes and wall panels in the zone of action of the burner. These water-cooled elements operate under super severe conditions. Moreover, the closer to the bath surface, the more severe the conditions. The blow-back of the oxygen jets reflected from the scrap lumps are the main cause of damage of the boxes and panels in the burner zone. Alternating operating modes of the burner reduces this problem, but does not eliminate it completely. In order to increase the durability of the water-cooled elements of the modules, some companies prefer to install them at a greater height, even though this installation increases considerably the length of oxygen jets and reduces the effectiveness of the bath blowing.

It should be emphasized that all the aforesaid concerning limited possibilities to heat scrap by burners of small power relates to the burners of jet modules as well. The need to increase the coverage of the burner on the scrap pile resulted in developing burners with a variable configuration of the flame. During the operation of these burners the shape of the flame could vary widely from the narrow round to a wide flat in a fan shape. Various options of such burners have been tested. However, considerable increase in their efficiency has not been obtained. A slight expansion of the flame coverage on front pile of scrap was compensated by a decrease in the depth of penetration of the flame into the mass of the scrap pieces due to the reduction of its kinetic energy.

2.2.3   Rotary Burners with Changing the Flame Direction

Another way to expand the area of the burner flame impact on the scrap was much more effective. The authors suggested replacing stationary burners with a constant direction of the flame on the rotary burner able to change the direction of the flame over a wide range during operation. The rotary burners have the following principle advantages. Moving flames from those already heated to the relatively cold zones of the scrap allows to increase by several times the effective power of the burners without shortening their operation time. High kinetic energy of the high-powered flames allows them to penetrate through the scrap pile down to the bottom. In this case, the heating gases pass the maximum distance in the layer of scrap, which considerably increases the heat transfer and the fuel efficiency coefficient. A quick and relatively uniform heating of large masses of scrap in the furnace freeboard can be provided by varying the number, location and power of rotary burners. The temperature of the flames of rotary oxy-gas burners has to be relatively low to prevent the intensive iron oxidation when high-temperature heating.

2.2.3.1 Slag Door and Oriel Rotary Burners

Two variants of oriel burners have been developed: for existing EAFs with an oriel tapping and for a new type of furnaces with an additional oriel, Fig. 2.2a, b. As for slag door burners, since great advantages of rotary burners over stationary ones were evident, they from the very beginning of their application were in most cases mounted on the brackets which allowed changing the flame-direction in the course of the heat. The first of these burners were hand operated and later the management was mechanized.

Fig. 2.2 HPR burners in the main and additional oriels of EAF (designations are given in the text). Patent of Russian Federation, No. 1838736 A3

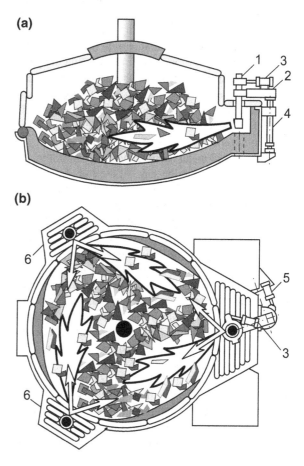

The installation of the burner in the existing oriel is shown in Fig. 2.2a. The water-cooled burner (1) is fixed to the bracket (2). It can be turned in both directions from the mid position around its axis with the help of the hydraulic cylinder (3). The gas and oxygen nozzles of the burner are located on its lateral surface near the lower end, at an angle to the bath. The vertical displacements of the burner are carried out with the help of the hydro-plunger (4) serving as a stand for the bracket (2) fixed to it in a manner allowing the rotation in a horizontal plane. The rotation of the bracket (2) with the burner is carried out with the help of the hydraulic cylinder (5). The cooling and shielding from the radiation when tapping are provided for all the mechanisms including the gas and oxygen lines of the burner.

The burner operates as follows. After charging of the first portion of scrap, the burner with the help of the mechanisms of horizontal and vertical displacement is inserted into the technological opening in the cover of the oriel intended for maintenance of the tap hole of the furnace. Then the burner is brought down into the chamber of the oriel to the lowest position, and the flame is ignited. Due to the fact that only insignificant amount of the scrap gets into the oriel chamber, it becomes possible to lower the burner almost to the level of the hot heel. Thereat, the gases pass the maximum distance in a pile scrap from the bottom up.

In the course of scrap heating, the burner is periodically turned around its axis from one end position to another with the help of the cylinder (3). By combining the heating of the layer of scrap from below and the periodic change of the direction of flame in the horizontal plane, the zone of flame action is being enlarged consid-erably, and the local overheating in the scrap are being eliminated. As a result, the optimum heat-transfer conditions are being ensured as well as, consequently, the possibility to increase sharply the power of the burner, the fuel efficiency and the medium mass temperature of scrap preheating.

As the liquid metal accumulates on the bottom, the burner is pulled up so that it does not immerse into the slag. At the end of the heating of the first portion of scrap, the burner is switched off and raised above the oriel cover. This way the nozzles of the burner are protected from clogging with splashed metal and slag with no use of a pilot flame. The operations of heating of each new portion of scrap charged are repeated in the same order. At the end of the heating of the last portion, the burner is raised to the upper position and is swung together with the bracket (2) in the horizontal plane to the off-position on the side of the oriel with the aid of hydro cylinder (5). The technological opening in the cover of the oriel is freed for con-ducting the maintenance of the tap hole of the furnace.

The use of one powerful oriel burner for heating the entire scrap charge is just enough to small furnaces only. The additional oriels in the form of special niches (6) in the side walls are required to install several such burners, Fig. 2.2b. These niches opened from the side of the bath make free space for the burner installation and for the changing of the flame direction. The number of niches and their positioning in the furnace can be different. The version with one of side niches intended for the installation of a burner is shown in Fig. 2.3. The lining of walls and bottom of the niche is the extension of the lining of the banks and the bottom of the furnace. From the top, the niche is closed with the water-cooled cover (1) with the

Fig. 2.3 Installation of HPR
burner in an additional oriel
(designations are given in the
text)

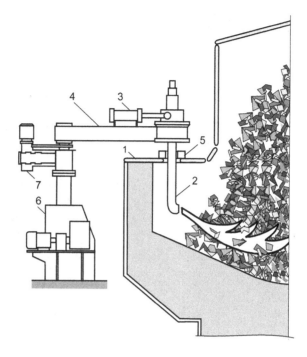

opening for inserting the burner (2). The burner is turned by using cylinder
(3) installed on the bracket (4).

   The application and the function of the mechanisms of each burner (2) are
completely analogous to those of the oriel burner shown in Fig. 2.2. The main
difference is that the shaft-type mechanisms with supporting rollers (6) installed on
the operating platform of the furnace are used for raising and lowering of the
burners. The burner displacement to off-position is carried out with the help of the
mechanism (7) turning the bracket (4) with the burner in a horizontal plane. This
brings the burner outside the boundaries of the furnace. The mechanisms other than
those already described can also be used in the installations of the oriel burners.

2.2.3.2   Roof Rotary Burners

In the second half of the eighties, arc furnaces in a number of Russian plants have
operated with the use of a metal charge contained a large amount of heavy rolling
trimmings. The high power rotary vertical roof burners called HPR-burners have
been developed by the authors for the high-temperature heating of scrap in such
furnaces. When charging the heavy scrap by two baskets there is a free space near
the sidewalls of the furnace enough for lowering such burners into the freeboard
through the roof ports near the roof ring, Fig. 2.4.

Fig. 2.4 Roof HPR burners
(designations are given in the
text)

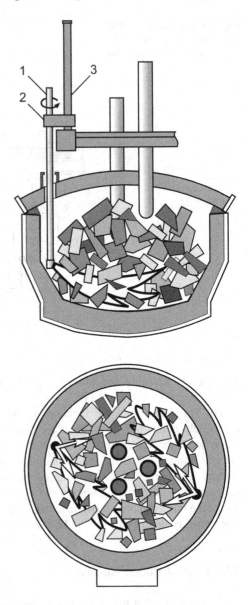

The roof burners (1) can be lowered and raised as well as turned around the
vertical axis up to 60° with the help of the mechanisms mounted on the carriage
(2) and on the column (3) along which the carriage moves. The gas and oxygen
nozzles are located on the side surface of the burner at an angle to the bath. The
flame direction changes within wide limits when the burner is moved along the
vertical axis or turned. Changing the flame direction of the burners in the course of
the melting period can be carried out either with the help of automatic device

according to the specified program, or manually. In the latter case, an operator directs flames into those zones of the freeboard where the scrap settlement is going slowly. This allows to consider the specifics of the heats and to accelerate the process essentially.

Prolonged industrial trials showed that two roof burners, when being properly controlled, ensure quite uniform heating of the entire scrap pile charged into the furnace. This is promoted by high kinetic energy of the flames of HPR-burners which allows them to penetrate deep into the layer of scrap practically reaching the bottom, as well as by the nozzle configuration providing the retarded mixing of gas and oxygen and reduced combustion temperatures. The burners are lowered into the freeboard only for the duration of their operation. This excludes a necessity to consume gas and oxygen for keeping up a pilot flame.

The tests of the burners were carried out in the old 100-t furnaces with the power of 32 MVA and in the 200-t furnaces with the power of 60 MVA. The furnaces had ceramic walls and roofs. The combined maximum power of two burners in the 100-t furnaces was 25 MW. Two roof burners, 15 MW each, and a slag door burner with the power of 5 MW were installed in the 200-t furnaces. The carried out experiments are of interest as the first experience of the commercially demonstrated technology for heating a scrap by HPR-burners the unit power of which exceeded that of burners currently used by more than tripled.

2.2.4 Two-Stage Scrap Melting. Industrial Testing of the Process

The amount of heat transferred to the scrap by the burners per unit time, i.e. the useful power of the burners P_{br}^*, is determined by the expression:

$$P_{br}^* = \eta_{gas} \cdot \Sigma P_{br} \tag{2.1}$$

η_{gas} gas efficiency coefficient
ΣP_{br} the total power of the burners, MW

The analogous expression for the useful electrical power of the arcs P_{el}^* is:

$$P_{el}^* = \eta_{el} \cdot P_{el} \tag{2.2}$$

η_{el} electrical energy efficiency coefficient
P_{el} electric power of the furnace, MW

Provided that the useful power of the HPR-burners P_{br}^* is close enough to the electrical power P_{el}^* there is a possibility to implement the so-called two-stage melting process. During the first stage the scrap is immediately after each charging heated only by the burners without the arcs, and at the subsequent second stage both the burners and the arcs or the arcs only are used. Such a process can provide not only the maximum possible reduction in electrical energy consumption due to heating a scrap by burners at the furnace freeboard but also shortening the total duration of the melting period. The latter will be able to take place if reduction in the duration of the second stage of the process $\Delta\tau_2$ exceeds the duration of the first stage τ_1. The tests of the two-stage process have showed that at the values of P_{br}^* close to those of P_{el}^* the use of HPR-burners allows realizing this condition.

2.2.4.1 Two-Stage Process in 100-t and 200-t EAFs

Parameters of these furnaces and HPR-burners installed on them are given above. The tests were carried out under the actual conditions of current production. The heats with and without the use of burners alternated which substantially eliminated the effect of the random factors. Scrap of each basket was heated and settled down first by the burners without the use of electrical energy, and then by the burners together with the arcs or by the arcs only. Furthermore, the burners were used during the power-off breaks caused by the reasons of organizational nature. During the simultaneous arcs and burners operation, the gas and oxygen consumption in the burners were being decreased.

The effectiveness of the burners operation was evaluated based on a change in the performance indices of the melting period, which, in comparison to the indices of the entire heat, are more closely correlated with the use of fuel for scrap heating. In the old furnaces which operated without treatment of steel in the ladle-furnace units, the electrical energy consumption for the heat and tap-to-tap time strongly depended on the technological factors not associated with heating of scrap by burners. In accordance with common practice, the end of the melting period was determined by formation of the flat bath with the temperature of approximately 1560–1580 °C.

Relative changes in the performance indices of the two-stage process in groups of the heats in the 100-t and 200-t furnaces did not significantly differ. The values of these indices averaged over all the heats are given in Table 2.1.

It should be noted that increasing the stability of arcing on the preheated scrap which was accompanied by an increase in average input power helped to reduce the melting period in the two-stage heats. At the test period of the 200-t furnaces the

Table 2.1 Performances of the two-stage heats

Average burners power as a percentage of average electric power for melting period	92.0%
Burners operation time without arcs as a percentage of melting period	32.5%
Shortening duration of melting period including power-on operation	9.4% 39.0%
Reduction in electrical energy consumption during melting period	37.5%
Gas energy efficiency coefficient η_{gas}	0.70
Average mass temperature of scrap heating by burners	700 °C

durability of the lining did not change and of the 100-t ones that even slightly increased. This indicates that the flames of HPR-burners despite their very high power are not dangerous for refractory and especially for water-cooled elements. Under conditions of the carried out tests with approximately the same power of the burners and arcs, HPR-burners were not inferior to the electric arcs with regards to energy efficiency, Table 2.1.

When testing these burners in the 100-t and 200-t EAFs, the main element of the fuel arc steel melting process namely the high-temperature scrap preheating by means of fuel has been first successfully implemented. Herewith, EAFs themselves acquired during the tests the basic features of the first fuel arc furnaces (FAFs). This was made possible thanks to the low power of transformers and the long tap-to-tap times at the old furnaces. In modern high-power EAFs with very short tap-to-tap time, it is not possible to carry out high-temperature scrap preheating by using fuel in the furnace freeboard as this would require increasing the burners' power to virtually unacceptable values. Nevertheless, the two-stage process is of practical interest for small EAFs with a low transformer power. This is confirmed by the results of testing the two-stage process for 6-t and 12-t plasma furnaces.

2.2.4.2 Two-Stage Process in Plasma Furnaces

Plasma furnaces were utilized in the production of special high-alloy steels and alloys by the method of remelting of clean materials in the neutral atmosphere. To test the two-stage process on these furnaces the slag doors were equipped with two shutters: one regular shutter for insulating the freeboard and the second additional shutter with a rotary oxy-gas burner installed in it. The second shutter had an exit aperture for the combustion products. In order to reduce the flame temperature the burners with a retarded mixing of gas and oxygen are used.

The tests were carried out during the charge melting period which was divided into two stages. At the first stage, immediately after charging of the entire metal-charge, the shutter with the burner was placed in the door of the furnace, and the charge was heated up to the maximum temperature. Then the burner was switched off, the shutters were transposed, the furnace was filled with argon, the plasmatron was switched on, and then the melting and the subsequent heating of the liquid metal were conducted as per conventional procedure without any changes in the electrical regime.

The design of the shutter with the burner allowed to change the flame direction within wide limits in the course of heating the charge and, at the same time, practically completely eliminated the air infiltration into the furnace. With the small dimensions of the bath, the burner ensured the sufficiently uniform heating of the entire charge. The two-stage heats and the regular heats without the use of the burner alternated. This decreased the effect of the random factors.

The entire charge was thoroughly weighed. This allowed to establish with high accuracy that the total oxidation of metal-charge in the two-stage process did not increase despite the presence in the charge of easily oxidized alloying elements.

Table 2.2 Performances of regular (A) and two-stage (B) heats on plasma furnaces

Performances	A	B
Power-on operation time, min	117	52
Electrical energy consumption, kWh/t	840	380

This result is explained by the uniform heating of the charge by the rotary burner as well as by to practically complete absence of free oxygen in the furnace freeboard.

An average burner power during the melting period was 5.5 MW with the natural gas energy efficiency coefficient of η_g equal to 0.5. Relevant parameters of the plasmatron were 2.3 MW and 0.45 respectively. Thus, in this case the burner excels electrical energy source with regard to both power and energy efficiency. The burner operation time τ_{br} was varied within 30 min. At τ_{br} = 25 min gas and oxygen flow rates were 43 and 86 m^3/t respectively and average mass temperature of preheating the charge has reached of 1050 °C. The enthalpy of the charge 200 KWh/t corresponds to this temperature. The enthalpy of the completely melted metal, in this case, is equal to 380 kWh/t. Therefore, during the two-stage heats approximately half of the total heat required for complete melting was obtained by the charge from the oxy-gas flame. The main results obtained in the tests on the two-stage process of plasma furnaces are shown in Table 2.2. In two-stage process the furnace power-on operation time and electrical energy consumption are reduced by a factor of 2.2.

The tests conducted revealed another important feature of the two-stage process. Despite a very high unit power of oxy-gas burners which more than twice exceeded the power of plasmatrons, the temperature of the furnace lining only slightly exceeded the temperature of the surface of the scrap and was not higher than 1450–1550 °C. With the arc heating, this temperature was approximately 200 °C higher. This difference is explained by the fact that the temperature of the oxy-gas flame is considerably lower than the temperature of the arc plasma, as well as by the fundamental differences in the laws of heat transfer by radiation and by convection. If the two-stage process is implemented in the EAFs, this feature ensures the decrease in heat losses due to water cooling the wall and roof panels of the freeboard.

In small EAFs, installing the oxy-gas burners whose power far exceeds the transformer power is not too much difficulty. For such furnaces, the established advantages of the two-stage process make it very promising.

2.2.5 Twin-Shell EAFs

Twin-shell furnaces are the unit consisting of two furnaces placed next to each other, Fig. 2.5, with common furnace transformer (1) connected by a secondary electrical circuit with a rotary system current conducting arms (2) and with

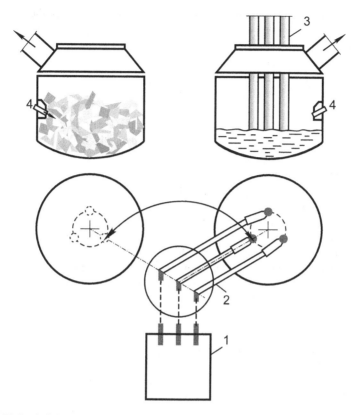

Fig. 2.5 Twin-shell furnace (designations are given in the text)

electrodes (3). This system allows to place electrodes alternately in each of the baths. Several similar furnaces were built in a number of countries, sometimes for the purpose of implementation of the special technological processes such as the units combining the functions of EAF and oxygen converter. This type of units is not discussed here. Both baths of the twin-shell furnaces were equipped with burners for heating of scrap and with devices for oxygen and carbon injection (4). The burners are also required to maintain the temperature of the hot heel at a sufficiently high level during periods of power-off operation.

Technological operations in the twin-shell furnaces are carried out in a certain order. For simplicity sake, one might discuss, as an optimal version, the operation of furnaces with single scrap charging. Charging by two baskets does not change the principle of their operation. While in the first bath during τ all the main technological operations with power-on such as melting preheated scrap, decarburization, heating the melt to a tapping temperature, and tapping itself are carried out, in the second bath for the same time τ_1 all the preparatory power-off operations are implemented namely the operations of closing a tap hole, scrap charging, and scrap preheating by burners.

On completion of the tapping the electrodes are transferred from the first bath to the second one the function of which becomes implementing all the operations with power-on during the time of τ. In this way, the tappings of the heats on the twin-shell furnace occur at regular intervals equal to τ. Maintaining the total duration of all the power-off operations at the same level with that of power-on operations is easily achieved by controlling the scrap heating time which is defined on leftovers.

The prime advantage of twin-shell furnaces is an increase in productivity due to implementing all the preparatory operations at a separate bath in parallel to the basic process operations. This provides considerable reduction in the interval between tappings since the duration of basic power-on operations is significantly shorter than tap-to-tap time of an equivalent single shell furnace.

The second important advantage of twin-shell furnaces in comparison with single shell ones is the presence of a quite long period of time to preheat scrap by burners as the total duration of the rest of preparatory operations is significantly shorter than τ_1. The scrap preheating reduces electrical energy consumption.

A significant rise in the cost and complexity of the design as well as an increase in maintenance costs are substantial shortcomings of twin-shell furnaces which interfered with their spread.

2.2.5.1 Twin-Shell Shaft Furnaces

As it noted earlier, heating the pile of scrap filling the furnace freeboard by means of sidewall burners had little effect because of non-uniformity of such heating. The most effective device for heating the scrap is a shaft. The highest average mass temperatures of scrap heating as well as the highest values of the gas heat efficiency coefficient close to 0.7 can be achieved when passing the heating gases through the layer of scrap filling the shaft. These considerations resulted in a development of twin-shell shaft furnaces.

In these furnaces, shafts through which scrap is charged into a furnace freeboard are located over roofs of each of bathes. The shafts have no devices retaining the scrap in them. Therefore, after charging a part of the scrap is eccentrically placed in the freeboard. The rest of the scrap fills the shaft.

Burners in twin-shell shaft furnaces can be located in both sidewalls of the freeboard and shafts themselves. Herewith their quantity and total power increase. All the products of combustion are discharged through the shaft which substantially increases the efficiency of the use of burners and provides scrap preheating up to an average mass temperature of about 450 °C. This temperature depends on the power of burners. In Table 2.3 the performances of the recently modified twin-shell shaft furnace are compared with those of one of the very powerful EAFs over the world.

The data in Table 2.3 leads to the following conclusions. Very high (almost the same with the EAF) productivity is achieved in the twin-shell shaft furnace at much lower values of electrical power and the tapping weight (by 1.6 times). Electrical energy consumption decreases by 20%. At identical power of burners installed in

Table 2.3 Performances of 200-t twin-shell shaft furnace and 320-t EAF

Performances	Shaft furnace	EAF
Tapping weight, t	200	320
Transformer power, MVA	140	240
Interval between tappings, min	36	56
Hourly productivity, t/h	325	343
Electrical energy consumption, kWh/t	290	362
Gas flow rate, m^3/t	11.7	3.6
Burners power of each bath, MW	32	32

these furnaces this indicates a considerable increase in their efficiency in the shaft furnace. It could be said that the twin-shell shaft furnaces would be promising if it would not be possible to obtain similar or even much better results in the single shell furnaces as well.

2.3 EAF with Preheating a Scrap by Off-Gases and Melting of Preheated Scrap in Liquid Metal

2.3.1 Conveyor Furnaces of Consteel-Type

Due to persistent efforts of Tenova company, conveyor electric arc furnaces Consteel designed by J.A. Vallomy have become considerably widespread. On the late 2013, there were over 40 of these furnaces worldwide. This steelmaking unit combines an electric arc furnace and vibratory conveyor (1) consisting of chutes made of sheet steel. A conveyor is adjacent to a side wall window of an arc furnace from the side opposite to a furnace transformer, Fig. 2.6.

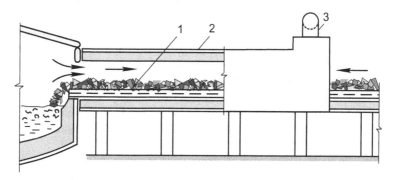

Fig. 2.6 Consteel furnace (designations are given in the text)

Total length of the conveyor is about 100 m. A part of the conveyor is inside of a refractory-lined tunnel (2) adjoining the furnace. The length of this tunnel is about 30 m. The off-gases leaving the furnace are removed through this tunnel. In the tunnel, the gases move in the opposite direction to the scrap and heat it up. Then the gases are directed through the water-cooled duct (3) into the bag filters. The conveyor chutes located in the tunnel are cooled by water.

Scrap charging into a bath is carried out by special water-cooled chute installed at the end of the conveyor. This chute is filled with scrap at regular intervals, then moved into a freeboard through a window, and the scrap is dropped into a bath. The width of the conveyor depends on furnace capacity and can reach 2.5 m. A furnace roof is opened only during the first heat after a furnace scheduled maintenance before which all liquid metal and slag is tapped from the furnace. During this heat, one basket of scrap is charged into the furnace from the top in order to accumulate on the bottom a sufficient amount of liquid metal to start continuous scrap charging by conveyor. During all other heats, the furnace operates with a hot heel and charging is carried out by conveyor only without opening the roof.

The rate of scrap charging into the bath is always kept equal to the rate of scrap melting in liquid metal. As a result, the bath remains flat during the whole melting process. During this period, the temperature of metal is kept at a constant level of 1560–1580 °C. Temperature increase above this level is not recommended in order to avoid sharp drop in durability of the refractory lining of the bottom and the banks of the furnace.

Long experience of operating Consteel-type furnaces showed that the mode of operation with virtually continuous scrap charging into flat bath and melting it in the liquid metal has a number of essential potential advantages such as energetical, technological, and ecological. Let us outline advantages of this mode of operation along with the factors which allow achieving these advantages:

- Shortening power-off furnace operation time and reduction of electric energy consumption. The contributing factors are as follows: elimination of time spent on opening and closing the roof for scrap charging; elimination of heat losses when the roof is open; improvement in conditions for stable maintenance of the foamed slag layer of the optimum thickness at the flat bath; increase in electric energy efficiency due to complete submerging of arcs into the foamed slag during practically the entire heat; the usage of a lined tunnel for scrap preheating by off-gases.
- Reduction of electrode consumption. In the absence of tall scrap column in freeboard, the risk of electrode damage is drastically lower; decrease of the freeboard height, especially in comparison with single charge furnaces, leads to reduction of electrode length and their side surface area, oxidation of which by furnace gases is one of the major factors of electrode wear.
- Increase in yield. The factors affecting yield increase are an absence of scrap oxidation by furnace gases and by flames of the burners, as well as reduction of FeO in slag and of oxidation level of liquid metal; furthermore, as off-gases pass through a conveyor tunnel, the large-size particles of dust containing iron

precipitate and return to a bath with the scrap; also, evaporation of metal does not occur, since the arcs do not contact solid scrap.

- Cost saving on electrical power supply and lowering of requirements for electrical grids. Due to elimination of the stage of unsteady burning of electric arcs during forming of cave in a column of scrap, there is reduction in voltage and frequency fluctuations generated by the arc furnaces in electric grids; this reduces costs on suppression of electrical interference in the grids as well as simplifies electric power supply to the furnaces.
- Increase of durability and service life of the sidewall and roof panels and other water-cooled elements. This is achieved by elimination of direct radiation of electric arcs which are being submerged in foamed slag at all times; simultaneously the risk of dangerous water leaks into the freeboard resulted from burnouts of the elements is significantly reduced.
- Reduction of gas emissions into the atmosphere and of expenditures on capture and purification of the off-gases; noise level reduction. These ecological advantages are achieved due to elimination of non-controlled dust and gas emissions which take place in case of open roof and increase the quantity of cleaned gases by two or three times; and also due to steady low-noise arcing in the foamed slag.

Along with these advantages Consteel furnaces have substantial drawbacks. Low productivity in comparison with that of modern EAFs of identical capacity is main one of them. Relatively slow melting of the scrap in the liquid metal which is the bottleneck of the whole process is a cause of decreasing productivity.

The rate of scrap melting in liquid metal is determined by intensity of convective heat transfer from metal to scrap which depends first and foremost on speed of metal streams flowing over the surface of scrap pieces as well as on difference in temperatures between liquid metal and scrap. Amount of heat Q transferred from liquid metal to scrap per unit time is defined by the following equation[1]:

$$Q = \alpha \times F \times \Delta t \qquad (2.3)$$

α, kW/(m^2 °C)   coefficient of convective heat transfer from liquid metal to scrap

Δt, °C   average difference in temperatures between metal and scrap pieces, per melting period

F, m^2/t   specific surface area of all the scrap pieces submerged into liquid metal

In Consteel furnaces, electric arcs are constantly submerged into foamed slag. They do not have direct heat contact with scrap and do not affect intensity of heat transfer from metal to scrap. The arcs participate in scrap melting only indirectly by maintaining metal temperature at a required constant level. Given a furnace capacity, such melting mechanism excludes the possibility of increasing melting

---

[1]This process is examined in detail in Chap. 3, Sect. 3.1.

rate by increasing the power of arcs. The latter must strictly correspond to the scrap melting rate which does not depend on the power of arcs. Thus, the power of arcs does not determine scrap melting rate, but, on the contrary, scrap melting rate determines the maximum permissible power of arcs which must not result in overheating of metal. In the EAFs, such a limitation of the electric power does not exist. As the arc power rises, both melting rate and hourly productivity of the furnaces increase.

In the modern EAFs, the scrap melting process may be divided into two periods. During the first period, the main body of scrap pile is melted down with the electric arcs in the furnace freeboard above the hot heel surface. Plasma of the arcs has a temperature of more than 5000 °C and possesses high kinetic energy. The heat energy of the arcs is transferred to the scrap by both radiation and convection. The intensity of these heat transfer processes is much higher than that in the liquid metal where the difference between a temperature of the liquid metal and scrap melting temperature amounts to 30–40 °C only. The power of the arcs is also higher than that of the Consteel furnaces. Furthermore, in EAFs the oxy-gas burners and the process gases formed during post-combustion of CO in the freeboard contribute to the heating and melting of scrap. At the first period of the scrap melting process in EAFs, all these factors ensure the higher melting rate in comparison with that in Consteel furnaces.

During the second period, upon completion of both forming of the flat bath and submerging of the arcs into a foamed slag, the rest of the scrap melts down in the liquid metal. If any, the mechanism of heat transfer to scrap is the same as in the Consteel process. However, just like in the first period, the rate of melting is considerably higher. It could be explained by the fact that the scrap lumps before submerging into the melt are heated in the EAF freeboard up to a temperature which is much higher than that in the tunnel of Consteel furnaces. Only a small portion of fine scrap placed on the bottom of the basket submerges into the hot heel not being preheated.

An increase in melting rate results in the higher productivity of the moderns EAFs as compared with the conveyor furnaces. This consideration is confirmed by actual data. For the entire range of the furnace capacities from 100 up to 350 tons, the rate of scrap melting in the EAFs is higher than that in the conveyor furnaces by approximately 1.6 times. Although the Consteel process eliminates times expended on upper scrap charging with baskets it does not compensate for such a strong lag of conveyor furnaces behind EAFs in the scrap melting rate. As a result, at identical capacity the hourly productivity of modern EAFs exceeds that of Consteel furnaces by an average of 20–25% [1].

Electrical energy consumption in the conveyor furnaces is approximately the same as or even higher than that of the EAFs operating without scrap preheating [1]. Electrical energy consumption is closely related to scrap preheating temperature. When Consteel furnaces were being developed, it was expected that scrap on conveyor would be preheated by off-gases up to 700–900 °C. Such preheating could have ensured quite considerable electrical energy savings. However, these expectations were not realized. Scrap preheating in Consteel furnaces proved to be ineffective. Mass average temperature of scrap preheating does not exceed 250 °C on the medium capacity furnaces and is even lower on the 350-t furnaces because of the increased scrap layer thickness (reaching 900 mm). According to energy balance of one of such furnaces, the enthalpy (heat content) of the preheated scrap is 16.4 kWh per ton of scrap. Interpolating between the values of temperature t_s equal to 100 and 150 °C, Table 1.1, Chap. 1, we find that the enthalpy of 16.4 kWh/t corresponds to the temperature of scrap $t_s = 125$ °C. Such low average mass temperatures of scrap preheating in the Consteel furnaces is explained by an unsatisfactory regime of heat transfer from off-gases to scrap in the lined tunnel. The off-gases pass at low rate along the surface of the scrap lying on the chutes. These do not penetrate into the inner layers of the scrap. As a result, only the top scrap layer, which receives heat also due to radiation from the lining, is heated intensively.

In countries like Russia, Norway, etc., operation of the Consteel furnaces during winter time causes considerable difficulties resulting from low scrap preheating temperatures. Getting into conveyor, snow and ice melt in the tunnel heated by gases. Water formed as a result does not have time to evaporate. It flows into the lower part of the conveyor chutes, mixes up with mineral debris contained in the scrap forming mud deposits and along with wet scrap are charged into the bath, which is not safe.

Reduced productivity and the absence of significant electrical energy savings are due mainly to the low scrap preheating temperature. The vibration conveyor does not meet the requirements for high-temperature heating of scrap. All the aforesaid does not allow to consider Consteel furnaces as promising aggregates capable of replacing modern EAFs.

It should be emphasized that the structural disadvantages of Consteel furnaces do not detract from the tremendous significance of the main achievement of J.A. Vallomy and specialists of Tenova as well. They have created a brand new process in EAF and proven the advantages of this process in practice. This process transfers the melting of scrap from the furnace freeboard to a liquid bath which means a new direction of development of EAFs that has a great future.

2.3.2 Shaft Furnaces with Fingers Retaining Scrap

The development of shaft furnaces is associated with the name of G. Fuchs. In the first 90-t shaft furnace the water-cooled shaft installed above the furnace roof did not have fingers retaining the scrap in the shaft. The scrap was charged into the furnace through the shaft. While the lower part of the scrap pile was located on the bottom of the furnace, its upper part was in the shaft. Gases from the furnace were evacuated through the shaft and heated the scrap located in it. As the scrap melted in the furnace, the entire scrap pile settled down. This created the free space in the shaft, which allowed charging of additional portions of the scrap.

Later G. Fuchs has developed and put into operation at plants of a number of countries the furnaces with one row of fingers in the lower part of a shaft, Fig. 2.7.

Fig. 2.7 Shaft furnace with
scrap retaining fingers

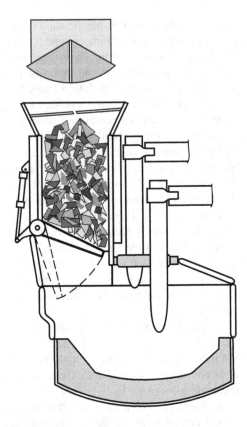

The scrap is charged into the furnace by two baskets. At the tapping the scrap of the
first basket heated by the off-gases during the previous heat lies on the fingers in the
shaft. After the tapping the fingers split apart, and the heated scrap is charged into
the furnace. After that, the cold scrap from the second basket is charged into the
empty shaft. The share of the scrap from the second basket remaining in the shaft is
heated by the off-gases passing through the shaft. As the scrap melts in the furnace,
the scrap in the shaft rapidly caves in and the shaft clears. This allows shutting the
fingers and charging the first basket of scrap on them for the following heat. By the
tapping time this portion of scrap is already preheated. With this heating method the
overheating and even the partial melting of the lower scrap layer do not create any
problems. The melt and the liquid slag formed flow down into the furnace and do
not obstruct splitting fingers apart for scrap discharging.

Heating a scrap on the fingers with gases passing through the layer upwards is
much more efficient than surface heating a scrap on a Consteel conveyor. This
contributes to increasing the average mass temperature of scrap heating and
reducing electrical energy consumption. Another advantage of shaft furnaces is that
a substantial part of the dust carried out from the freeboard settles down in the layer
of scrap. Due to this fact the yield is increased by approximately 1%. Typical
performances of shaft furnaces with fingers retaining scrap are shown in Table 2.4.

Table 2.4 Performances of shaft furnace with scrap retaining fingers	Tapping weight, t	135
	Tap-to-tap time, min	38
	Transformer power, MVA	120
	Productivity, t/h	214
	Minimum electrical energy consumption, kWh/t	285

There is no reliable data on average mass temperatures of scrap preheating for the finger shaft furnaces. It is known only that the smallest pieces in the lower layer of scrap on the fingers are melted down and fall into the bath in the form of a steel rain. However, this temperature with sufficient accuracy for practical purposes can be estimated by calculation.

2.3.2.1   Calculation

Let us consider the heat balance of the metal bath, namely the metal bath, and not the balance of the entire bath including slag. When melting down a preheated scrap its heat energy is completely absorbed by the liquid metal. Therefore, the average mass temperature of preheated scrap t_S and reduction in electrical energy consumption ΔE_{EL} obtained due to scrap preheating are connected a relationship close to unique:

$$\Delta E_{EL} = c_s \cdot t_s / \eta_{EL} \qquad (2.4)$$

c_s average calorific capacity, kWh/(kg °C)
η_{EL} the coefficient of electrical energy efficiency for heating a liquid metal

The product of $c_s \cdot t_S$ is enthalpy of scrap ΔE_{EL}, kWh/t, see Chap. 1, Table 1.1. The η_{EL} coefficient is determined by the following expression:

$$\eta_{EL} = \eta_{SEC.C} \cdot \eta_{ARC} \qquad (2.5)$$

$\eta_{SEC.C}$ the coefficient considering electrical energy losses of transformer and the secondary electrical circuit including electrodes
η_{ARC} the heat energy efficiency coefficient of the arcs considering energy losses during the heat transfer from the arcs to the liquid metal.

For the shaft EAFs operating with the flat bath and with the arcs practically continuously immersed into a foamed slag it could be assumed that $\eta_{SEC.C} = 0.93$; $\eta_{ARC} = 0.90$; $\eta_{EL} = 0.93 \cdot 0.90 = 0.84$. Complete immersion of the arcs into the slag does not provide the η_{ARC} coefficient equal to 100%, as sometimes considered. The immersed arcs transfer the heat mostly to the slag, but the slag, though it is mixed with the metal, radiates a considerable portion of the absorbed heat into water-cooled sidewall and roof panels of the furnace. The aforesaid value of the

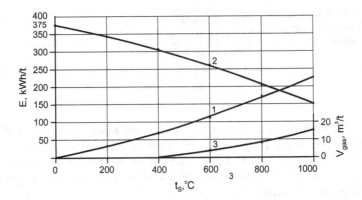

Fig. 2.8 Electrical energy consumption and gas flow rate vs average mass scrap preheating temperature t_S

η_{ARC} and, consequently, the η_{EL} as well should be considered rather somewhat overstated than understated. This fact is confirmed by the investigation of a goodly number of modern EAFs which had shown that during the melting the η_{EL} coefficient ranges from 0.6 to 0.8 [2]. In this investigation, as opposed to the others, it is stated that the η_{EL} coefficient applies to the metal.

Given η_{EL} = 0.84, the results of the calculation according to Eq. (2.4) are demonstrated by Fig. 2.8. As the scrap preheating temperature t_S raises, the reduction in electrical energy consumption ΔE_{EL} grows at the increasing rate, curve ΔE. At the t_S = 1000 °C, in comparison with the furnace operation without scrap preheating, the ΔE_{EL} reaches 225 kWh/t. Therefore, if we assume that without scrap preheating (t_S = 0) the electrical energy consumption EAF is on average of 375 kWh/t, then at the t_S = 1000 °C it will decrease to 150 kWh/t (375 −225 = 150). At the t_S = 800 °C, the ΔE_{EL} = 170 kWh/t and the electrical energy consumption amounts to 205 kWh/t (375−170 = 205). Such electrical energy consumptions are unachievable in EAFs operating without scrap preheating.

In operating shaft furnaces with fingers retaining scrap, the minimal electrical energy consumption amounts to about the 285 kWh/t, Table 2.4. Maximum values of ΔE equal to 90 kWh/t (375−285 = 90) and t_S = 450 °C, Fig. 2.8, correspond to this consumption. Thus, it can be assumed that, in accordance with the above calculation, the average mass temperature of scrap preheating does not exceed on average of 400 °C in the shaft furnaces with fingers retaining scrap, Fig. 2.7, when charging two baskets.

2.3.2.2 EAF Quantum

Later design of a finger shaft furnace was persistently improved. The latest variation of the furnace, named Quantum, developed by Primetals Technologies, Germany,

Fig. 2.9 Quantum shaft furnace (designations are given in the text)

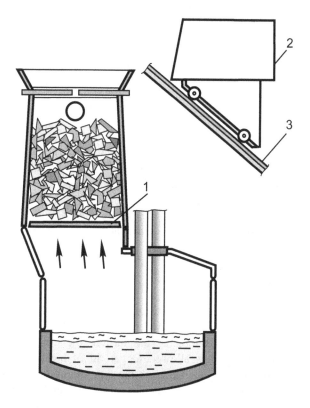

along with other innovations comprises a new system of shaft charging and improved design of fingers (1) retaining scrap, Fig. 2.9 [3].

Scrap is charged into the shaft from above by a tilting container (2) moving up and down on inclined elevator (3) rather than baskets with a crane. The container is loaded on scrap yard by the two hoppers. The loading system is automated. It allows covering the shaft with a hood and considerable decreasing uncontrolled gas-dust emissions from the shaft during the charging of scrap. The duration of the complete cycle of charging the next portion of scrap into the shaft is 5 min. This cycle includes the steps of: movement of a container with scrap up to the position of discharging, opening the shaft, scrap discharging, and the movement of the container down to the position of its load with scrap.

As opposite to the fingers shown in Fig. 2.7, in the EAF Quantum the fingers, similar to pitchforks, are introduced into the shaft through its sidewalls. To charge a batch of the preheated scrap into the bath the fingers are pulled out of the sidewalls of the shaft. After charging they are immediately introduced into the shaft to receipt the next batch of scrap. Thus, unlike the system shown in Fig. 2.7 all the scrap is heated on the fingers and not just the first portion. It was assumed that this would allow increasing the average mass temperature of the scrap.

Table 2.5 Basic performances of EAF Quantum [3]

Tapping weight, t	100
Tap-to-tap time, min	36
Transformer power, MVA	80
Productivity, t/h	167
Electrical energy consumption, kWh/t	280–300
Gas flow rate, m^3/t	4.4

The preheated scrap is charged into the liquid steel bath in several portions. Melting the scrap is realized in flat bath with all the advantages previously discussed, Sect. 2.3.1. The furnace is equipped with two vertical roof tuyeres for oxygen bath blowing.

Basic performances of the 100-ton EAF Quantum in Vepacrus (Mexico) for the first (2015) year of its operation are given in Table 2.5.

2.3.2.3 EAF SHARC

The main feature of a 100-ton shaft DC SHARC-type EAF operating in Hellenic Halyvourgia, Greece, is two shafts disposed symmetric about the central electrode [4]. During operation of the furnace the electrode diameter was increased to 700 mm which is allowed to get rid of breakdowns and lower electrode consumption to 0.6 kg/t. Bottom pin type electrode has a durability of 3500 heats.

The 34 fingers, which can be introduced and pulled together or separately, retain scrap in the shaft. Along with the symmetry of the entire construction such a manner of finger operation provides uniform distribution of scrap on the bath perimeter and uniform melting a scrap as well.

Due to the local conditions the furnace works only for 8 h on weekdays and 13 h on weekends. Because of such an intermittent work schedule the data on furnace productivity and on electrical energy consumption given in Table 2.6 should be considered as a referential data.

Table 2.6 Basic performances of SHARC EAF

Tapping weight, t	97
Tap-to-tap time, min	48
Transformer power, MVA/MW	78/54
Productivity, t/h	120
Electrical energy consumption, kWh/t	290
Gas flow rate, m^3/t	6.4

2.3.3   Shaft Furnaces with Pushers of the COSS-Type

The off-gas heat efficiency when heating a scrap in a shaft is higher by approximately three times than that when scrap is heated on a Consteel conveyor. On the other hand, continuous furnace operation with the flat bath is an advantage of the Consteel process. In order to combine these both advantages G. Fuchs has developed and implemented shaft furnaces with the practically continuous charging of scrap into the liquid bath. The design of such a furnace named COSS is schematically shown in Fig. 2.10.

A rectangular shaft (1) is installed on the cart (2) next to the furnace. The shaft is connected to the furnace with a short tunnel (3). A sliding gate (4) opens for charging of scrap into the shaft. The charging is carried out with the power-on and does not interrupt the furnace operation. A gas duct (5) is placed under the sliding gate (4). The shaft is lined with the massive steel segments and has no water-cooled elements which could be damaged during the scrap charging. The mass of scrap in the shaft is gauged by the measuring elements on which the shaft rests. This makes it possible to control the rate of charging of scrap into the furnace.

The scrap is charged continuously into the liquid bath in separate portions with the help of the pusher (6) which is moved forth and back by the well protected hydraulic cylinder. Such s charging in principle does not differ from the charging of scrap by a vibrating conveyor and allows maintaining a flat bath. Thus, all the

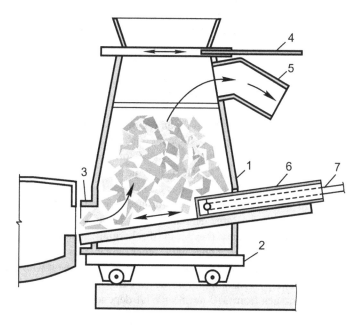

Fig. 2.10 COSS shaft furnace system (designations are given in the text)

above considered advantages of the operation with a flat bath proven by the experience of the Consteel furnaces operation are retained in COSS furnaces, Sect. 2.3.1.

The pusher discharges scrap from the bottom layer which has the highest temperature. Unlike the first finger shaft furnaces, in the COSS furnaces the off-gases heat all the scrap. This made it possible to expect the higher temperatures of heating. However, the data on the electrical energy consumption for the COSS furnaces does not confirm this expectation.

On three such furnaces of 140–150 t capacity with transformers of 80 MVA the minimum electrical energy consumptions were 300–400 kWh/t.[2] Such electrical energy consumptions correspond to the average mass scrap preheating temperatures of 350–400 °C, Fig. 2.8. Productivity of the furnaces ranged from 125 to 149 t/h, which was significantly lower than that of modern EAFs of the same capacity. This is explained not only by the relatively low melting rate of the scrap in the liquid metal at insufficiently high scrap preheating temperatures but also by the low electrical power of the furnaces.

The pictures of the scrap being discharged from the shaft of the COSS furnace into the window are presented by some publications. These pictures show that the temperature of the surface of scrap pieces at the discharging window can achieve about 700 or 800 °C. However, it should be kept in mind that in this zone the frontal surface of the front scrap layer is exposed to direct heat radiation from the furnace freeboard. Therefore, the mentioned temperatures are much higher than the actual average mass temperatures of the scrap located behind the window. If the actual average mass temperatures could reach 700–800 °C then the electrical energy consumption in these shaft furnaces would be close to rather the 200 kWh/t than to 300, Fig. 2.8.

2.3.3.1 Shaft Furnaces of ECOARC-Type

These furnaces have been developed in Japan regardless of the works of G. Fuchs. Four of these medium-capacity furnaces operate in Japan, one in Thailand, and one in South Korea. The structural scheme of these furnaces is close to the COSS scheme. The main feature of the ECOARC is that the shaft is attached to the furnace shell and forms with it a single whole. In this case, the scrap discharging window of the shaft is aligned with the window of the furnace sidewall without clearance.

In COSS-type furnaces, a large amount of air is infiltrated into the furnace through a gap between it and the scrap discharging window of the shaft. Elimination of this infiltration along with other advantages allowed reducing electrical energy consumption of ECOARC furnaces in comparison with the COSS furnaces. Electrical energy consumption of the 70-t ECOARC furnace of 50 MVA power in Thailand ranges from 310 to 285 kWh/t depending on the scrap quality.

---

[2]G. Fuchs's presentation Yekaterinburg, Russia, 2012.

The pusher of the ECOARC furnace is cooled by compressed air which excludes the possibility of high-temperature scrap preheating.

2.4   Factors Hindering Wide Spread of Shaft Furnaces

Shaft furnaces have been operating for more than 20 years in a number of countries around the world. During this time they were considerably improved but still not widely spread. This can be explained by the two factors. Firstly, shaft furnaces as well as Consteel furnaces are much inferior to modern EAFs in terms of productivity due to the long scrap melting time in liquid metal. The reasons for this, in common with the Consteel furnaces, were discussed in detail in Sect. 2.3.1.

Secondly, reduction in electrical energy consumption achieved in shaft furnaces does not compensate decreasing productivity and does not justify the application of the new complex equipment that requires additional maintenance. Of particular note is the fact that a minimal electrical energy consumption of 285 ± 15 kWh/t was almost the same at all shaft furnaces in spite of their rather significant design differences.

The energy consumption of 285 kWh/t has already been achieved 20 years ago at one of the first finger furnaces, although only part of the scrap was heated in the shaft of this furnace [5]. The furnaces of COSS-, Quantum-, SHARC-, and ECOARC-type created much later operate with the same electrical energy consumption. In order to reduce this consumption the scrap preheating to a temperature much more than 400–450 °C is needed. It was not possible to implement such preheating on any of the new shaft furnaces which operate with scrap heating only by off-gases. This leads to the conclusion that a common cause for all furnaces which prevented the further increase in the scrap preheating temperature and the corresponding reduction in electrical energy consumption is the insufficient specific heat power of the off-gas flow P_{gas}, kWh/(t·min). With a short tap-to-tap times this power is practically identical in all shaft furnaces. It is determined by approximately equal consumption of carbon-containing materials, kg/t of liquid steel, which does not depend on the furnace capacity.

Unfortunately, data on P_{gas} is absent. Instead of this key parameter the specific values of heat losses from off-gases Q_{loss}, kWh/t of steel, are given in many publications. According to the data from different authors, Q_{loss} ranges from 80 to 220 kWh/t. It does not specify to which time of power-on operation these values correspond. The value of Q_{loss} cannot substitute for P_{gas}, since it characterizes quantity of heat rather than power of heat flow. Using the value of Q_{loss}, it is impossible to determine average-mass scrap temperature t_S which can be obtained in case of scrap heating by off-gases in a given heating time τ. Determination of scrap preheating temperature t_S requires calculation of the heat power of the off-gas flow P_{gas}.

2.4.1 Calculation of the Maximum Values of the Power of the Heat Flow of Off-Gases and Temperature of Scrap Heating by These Gases in the Shaft

Let us consider the operation of a shaft furnace with flat bath under the following conditions. The only carbon source mainly determining the heat power of the flow of off-gases is coke charged into the liquid bath. The lumpy coke is charged via the opening in the furnace roof. The powder coke is injected into the bath by injectors. The carbon content in the coke is 80%. Carbon contained in scrap is not considered as its concentration does not exceed 0.20%. Further calculations are carried out per a ton of liquid steel.

The total consumption of coke M_{cok} is taken equal to 14 kg/t which are close to maximum consumption used in practice. The total consumption of carbon $M_C = 18 \times 0.8 = 14.4$ kg/t. It is assumed that carbon contained in the coke is oxidized to CO with oxygen injected into the bath with an intensity of $J = 0.9$ m^3/(t·min). In modern EAFs such intensity is close to maximum. The intensity of CO evolution from the bath (CO flow rate) V_{CO} is determined by the equation: $V_{CO} = 2$ k·J, m^3/(t·min). The coefficient k = 0.7 takes into account that some of the injected oxygen is consumed in the oxidation of iron,[3] silicon and manganese, and that a certain amount of O_2 is not absorbed by the bath. The coefficient 2 corresponds to the reaction equation: $2C + O_2 = 2CO$.

$V_{CO} = 2 \times 0.7 \times 0.9 = 1.26$ m^3/(t·min). Knowing the amount of carbon M_C in the metal bath and the intensity of CO evolution from the bath V_{CO}, both the blowing time τ required for carbon removal and equal to it the heating time of scrap in the shaft can be determined.

The flow of CO in volume units $V_{CO} = 1.26$ m^3/(t·min) corresponds to a flow in mass units $M_{CO} = 1.26 \times 1.25 = 1.57$ kg/(t·min), where 1.25 is density of CO, ρ kg/m^3. According to the reaction equation $C + 0.5\,O_2 = CO$, one kilogram of CO is formed from 0.428 kg of C. To the flow of CO equal to 1.57 kg/(t·min) corresponds the flow of carbon $M_C = 1.57 \times 0.428 = 0.672$ kg/(t·min). The amount of carbon removed from the bath is 14.4 kg. The time of bath blowing $\tau = 14.4/0.672 = 21$ min. The time of scrap heating in the shaft is the same.

Further it is assumed that all CO evolving from the bath is sucked into the shaft and only there burns to form CO_2. Thus, it is assumed that the off-gases consist of CO only, and that all the chemical energy from CO post-combustion evolves in the shaft. These operating conditions of the furnace are idealized. They create the most favor opportunities for reaching the maximum temperature of scrap preheating t_S.

Carbon monoxide introduces both physical q_{ph} and chemical q_{ch} heat into the shaft with scrap.

---

[3]It is understood that a part of the resulting iron oxides is not reduced by carbon with evolution of CO and remains in the slag.

$q_{ph} = V_{CO} \times c \times t$, kJ/(t·min), t = 1570 °C is temperature of metal during the stage of scrap melting in the furnaces with the flat bath, c = 1.475 kJ/(m³ °C) is heat capacity of CO at t = 1570 °C. q_{ph} = 1.26 × 1.475 × 1570 = 2918 kJ/(t·min) or q_{ph} = 0.81 kWh/(t·min).

The thermal effect of the reaction $CO + 0.5O_2 = CO_2$ is 3.51 kWh/m³ CO. q_{ch} = 1.26 × 3.51 = 4.42 kWh/(t·min). Heat power of the flow is $P_{CO} = q_{ph}$ +q_{ch} = 0.81 + 4.42 = 5.23 kWh/(t·min).

Let us determine the quantity of heat Q introduced into the shaft during τ = 21 min: Q = 5.23 × 21 × 0.9 = 98.8 kWh/t of scrap, where 0.9 is the yield of liquid steel per 1 ton of scrap. With the heat efficiency coefficient of gases in the shaft equal 0.7, the enthalpy (heat content) of scrap E_S will amount to: 98.8 × 0.7 = 69.2 kWh/t. The average-mass temperature of 460 °C corresponds to this enthalpy, Table 1.1. The result of this calculation is in good agreement with the aforesaid estimations of the scrap preheating temperatures in shaft furnaces according to electrical energy consumption. The actual values of t_S should be lower than the calculated ones since in real conditions a significant share of the CO burns even before entering the shaft and gives away some of its heat to the water-cooled panels.

The above calculation as well analysis of the accumulated experience of work and improvement of the EAFs leaves no doubt that the energy of the off-gases is in principle absolutely insufficient for high-temperature preheating of the scrap. To realize this most important element of a new technology an additional more powerful and efficient source of energy is needed. Natural gas should be used as such a source. Dependences of electrical energy consumption on the average mass temperature of scrap preheating were discussed in detail in Sect. 2.3.2.1. Now it is necessary to establish to what extent the increase in this temperature can accelerate the process of melting scrap in liquid metal and increase the productivity of the furnaces as well. Let us examine the results of studies of this process.

References

1. Tou[l]ouevski YN, Zinurov IY (2015) Electric arc furnace with flat bath. Achievements and Prospects, Springer, 32–35
2. Pfeifer H, Kirschen M, Simoes JP (2005) Thermodynamic analysis of EAF electrical energy demand. In: 8th European electric steelmaking conference, Birmingham
3. Apfel J, Mueller A, Beile H (2016) EAF Quantum—results of 2015. In: EEC 2016 Conference, Venice, Italy, Proceedings, index 47
4. Metzen A, Halyvourgia H, et al (2016) SHARC—The Cost Effective melting machine, EEC 2016 Conference, Venice, Italy, Proceedings, index
5. Hassig M, Fuchs G, Auer W (1999) Electric Arc furnace technology beyond the year 2000. MPT Int (1): 56–63

Chapter 3
Experimental Data on Melting a Scrap in Liquid Metal Required for Calculation of This Process

Abstract The process of scrap melting in iron-carbon melt is very complex. This process combines the following processes: convective heat transfer from the melt to scrap closely related to hydrodynamics of the liquid bath; heat transfer by thermal conductivity from the surface to internal layers of scrap pieces; and also diffusion processes of saturation of the surface of scrap pieces with carbon. Almost all of our knowledge in this field has been obtained experimentally in laboratory conditions. In modern electric arc furnaces, the content of carbon in hot heel is approximately the same as in scrap. Therefore, only those processes of scrap melting, where the rate is practically entirely determined by heat transfer conditions, are considered. One of the basic experimental methods used for study of melting processes is a method of melting samples immersed into the melt. The data, obtained by this method at McMaster University, Hamilton, Canada, after their analysis and additional calculations by the authors were used to determine the melting time of scrap in a liquid metal under real conditions of the EAF.

Keywords Melting scrap in iron-carbon melt · Methods of study · Immersion of scrap samples into liquid metal

3.1 Features of Scrap Melting Process

The liquid metal where the scrap is melted always contains a certain amount of carbon. The process of scrap melting in iron-carbon melt is very complex. This process combines the following processes: convective heat transfer from the melt to scrap closely related to hydrodynamics of the liquid bath; heat transfer by thermal conductivity from the surface to internal layers of scrap pieces; and also diffusion process of saturation of the surface of scrap pieces with carbon. Such a complex combination of thermal, hydrodynamic and diffusion processes practically does not lend itself to analytical methods of investigation. Almost all of our knowledge in this field has been obtained experimentally in laboratory and industrial conditions.

Y.N. Toulouevski and I.Y. Zinurov, *Fuel Arc Furnace (FAF) for Effective Scrap Melting*, SpringerBriefs in Applied Sciences and Technology, DOI 10.1007/978-981-10-5885-1_3

Let us consider the effect of carbon content in the melt on the melting process of scrap. Carbon is transferred from the melt to the scrap by diffusion with an intensity proportional to the difference in carbon concentrations in the melt and in the scrap. When increasing the concentration of carbon in the surface layer of scrap pieces the scrap melting temperature of this layer drops sharply and the scrap melting rate increases. At a low temperature of the melt and a high content of carbon in it, the scrap melting rate can be determined entirely by the diffusion process which becomes the basic one.

It is in such conditions that the scrap is melted at the initial stage of the process in oxygen converters. A temperature of hot metal poured into a convertor is lower than the melting point of low-carbon scrap by on average about 200 °C. Melting such a scrap in hot metal would be impossible without the carbon diffusion. When the surface of scrap pieces is saturated with carbon, its melting point falls below the temperature of hot metal which rapidly increases with oxygen bath blowing. The scrap, layer-by-layer, is converted into a liquid state and mixes with the liquid bath.

The term "melting" itself does not correspond fully to the nature of such a process inseparably associated with diffusion of carbon. On the other hand, the term "dissolving" is not quite applicable here since dissolving means mixing with the liquid bath through diffusion but without preliminary transition of solid metal into a liquid state. The term "diffusion melting" can be used when describing similar processes. Molecular diffusion is quite slow process in order to be able to serve as a basis for commercial modes of steelmaking. However, in converters this process is many-fold accelerated due to intensive bath stirring with oxygen blowing.

In modern electric arc furnaces making semi-product with subsequent out-of-furnace processing a metal the content of carbon in hot heel is much the same as in scrap. Under such conditions, diffusion melting cannot be noticeably developed.[1] Therefore, only those processes of scrap melting in EAFs, where the rate is practically entirely determined by heat transfer conditions, will be analyzed further.

When melting a scrap in the liquid metal the heat is transferred from metal to scrap by convection. Convective heat transfer (convection) occurs always when a fluid flows around a surface of a solid body. Heat can be transferred both from a liquid to a solid surface and in an opposite direction depending on liquid and surface temperatures. Convection is inseparably related to fluid motion. There is no convection in immovable medium.

Convection is a very complicated phenomenon. Nevertheless, for generalization of experimental data and uniformity of calculation methods, all various convection cases are calculated using the same simple equation:

$$Q = \alpha \left(t_L - t_C\right) \cdot F \tag{3.1}$$

---

[1]The exceptions are the furnaces operating with use of hot metal, which are not reviewed here.

This equation shows that quantity of heat Q transferred from a fluid to a surface of a solid body per unit of time is proportional to the area of this surface F and the difference between the temperature of the fluid t_L and that of the surface t_C. When scrap is melted then $t_L > t_C$. If $t_C > t_L$, heat is transferred from the surface to the fluid.

Coefficient of proportionality α, $W/(m^2 \, °C)$, is called coefficient of convection heat transfer. It describes the intensity of the heat transfer process. Equation (3.1) seems simple only at first glance. In fact, the entire complexity of convective heat transfer process and difficulties of its calculation are concealed in the only value, i.e. in coefficient of heat transfer α. This equation is used in calculations of convection processes which are very different by their physical nature. Experimental results are processed in such a way that allows determining the dependence of α on factors which determine conditions of the process run. In terms of their physical properties liquid metals including the melt of low-carbon steel are very different from other liquids. The intensity of the heat transfer processes in liquid steel, and consequently the values of α as well, are higher than those in water by several times.

When cold pieces of scrap are immersed into a liquid metal bath of electric arc furnaces or oxygen converters a solidified layer of metal is always formed at the surface of the pieces. At first the thickness of the solid layer increases then its complete melting down occurs and only then the melting of the scrap itself begins. The phenomenon of solidification considerably slows down the scrap melting process.

3.2   Studies of the Melting Process by the Method of Immersion of Samples in a Liquid Metal. Analysis of the Results

3.2.1   Melting of Single Samples of Scrap with a Solidified Layer and Without Solidifying

One of the basic and most frequently used experimental methods of study of melting processes is a method of melting samples immersed into the melt. The results obtained by this method at McMaster University, Hamilton, Canada [1] are of great practical interest. Cylindrical samples made of low-carbon steel 25, 32 and 38 mm in diameter were immersed in the crucible of induction furnace containing liquid metal. The length of immersed part of samples was 170 mm. The chemical composition of the metal was close to the composition of the samples. The initial temperature of samples and the duration of their immersion into metal were varied. At various fixed moments of time, the samples were taken out from the crucible, cooled in water, and weighed. The radius of the immersed part of the samples R was calculated using change in mass. Based on the data from these experiments, the curves of sample melting were plotted. A temperature of metal in the crucible was maintained at the level of 1650 ± 5 °C. During the immersion of the samples, the induction furnace was switched off so that forced stirring of the metal was absent. Oxidation of the samples in air and in the furnace for their preheating was prevented by the use of argon.

There are two possible alternatives of melting: with or without solidifying of metal layer on a sample. Realization of one or the other of these variants depends on the temperature of the metal and the initial temperature of the sample. These temperatures determine the value of the three different by their physical nature specific heat flows (kW/m^2) taking place in the process of melting. First of them q_1 is the convective heat flow from liquid metal to sample surface or to solidified layer, Fig. 3.1. This flow is determined by Eq. (3.1).

The heat flow q_2 occurs due to thermal conductivity of the layer and the sample. It is directed from the surface of the sample toward its axis; it is consumed to increase the temperature of the sample, and is determined by the equation:

$$q_2 = \lambda \cdot \frac{dT}{dR} \tag{3.2}$$

λ coefficient of thermal conductivity of layer and sample, W/(m^2 °C)
$\frac{dT}{dR}$ temperature gradient on sample surface

The heat flow q_3 is related to the conversion of metal from liquid into solid state or vice versa at constant temperature t_C. This flow is determined by the equation:

$$q_3 = H \cdot \rho \cdot v \tag{3.3}$$

H = 75 W h/kg   latent heat of melting (solidifying) of low-carbon steel
ρ = 7900 kg/m^3 density of solid metal
V   speed of interface migration between solid and liquid metal, m/h.

In case of solidifying, this interface moves to the left, and H has a plus sign, Fig. 3.1. The thickness of solidified layer increases, and heat is emitted. During the process of melting, the interface travels to the right, the thickness of layer decreases, H changes its sign to minus, and heat is absorbed. For the cylindrical sample, $v = \frac{dR}{d\tau}$.

Let us examine the most important for practice case: melting with solidifying. It is always takes place when cold scrap is immersed in metal with temperature of 1550–1650 °C which is common for EAFs and converters. A typical curve of melting cold (0 °C) sample of 25.4 mm diameter in metal at temperature 1650 °C is shown in Fig. 3.2 [1]. The process of melting with solidifying consists of two stages. During the first stage with duration of τ_C, metal layer is being solidified onto the sample. The thickness of this layer first increases, reaches maximum at point m,

Fig. 3.1 Schematic diagram of temperature profile of steel bath and sample immersed into it. *1* – liquid metal surface; *2* – solidified layer; *3* – steel sample (designations are given in the text)

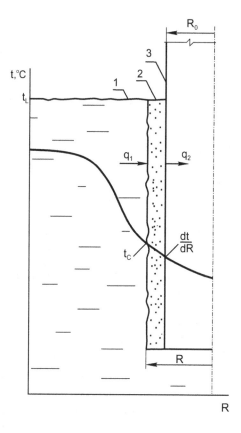

Fig. 3.2 Dependence of relative sample radius, R/R_0, on an increase in immersion time of sample into the melt, τ [1]. τ_C—period of metal solidifying onto sample and melting of solidified layer; τ_L—period of melting a sample itself; d = 25.4 mm; t_L = 1650 °C (designations are given in the text)

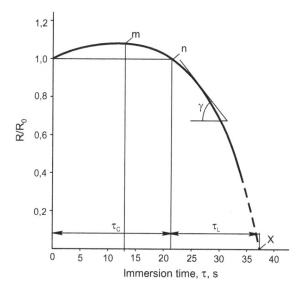

after which the layer begins to melt and disappears at point n, Fig. 3.2. After melting

of the layer is completed, melting of the sample starts. This second stage with duration of $_L$ ends with complete melting of the sample at point x. Total melting time $= _C + _L$ in this case is 37.5 s, Fig. 3.2.

To better understand mechanism of melting of scrap in liquid metal as well as to determine the effectiveness of various methods of intensification of this process, a detailed analysis of the shape of the melting curve, Fig. 3.2, is needed and, in particular, of presence of maximum of this curve at point m. Due to quite high value of α in case of heat transfer in liquid metal, the temperature of sample surface right after immersion increases very rapidly to melting (solidifying) point of the metal, t_C. The heat flow to the sample surface q_1 is stabilized at the value determined by Eq. (3.1). Since the values of α, t_L, and t_C do not change further, the flow q_1 remains constant during the whole process of sample melting. At the same time, a thin metal layer contacting the sample surface cools down and solidifies at the same temperature t_C.[2] Following the first layer, the second, third, and so on layers also solidify. The thickness of the layer solidified on the sample increases. The temperature on its surface stays constant and is equal to t_C.

The heat from the solidified layer transfers to the sample due to thermal conductivity (flow q_2), Fig. 3.2. At the first moment after immersion of the sample, both gradient of temperature on the sample surface $\frac{dt}{dR}$ and flow q_2 have maximum values, Eq. (3.2). Then, as the temperature of sample rises, the gradient $\frac{dt}{dR}$ and the flow q_2 are reduced. However, due to high thermal conductivity of the sample, the condition $q_2 > q_1$ remains unchanged for a certain period of time.

During this period, the intensity of metal cooling by the sample is gradually reduced. The thickness of the solidified layer continues to increase; but the rate of this increase gradually drops (melting curve in the interval from the beginning of the process to point m, Fig. 3.2. The heat balance of the metal—sample system, for which heat input is required to be equal to heat output, is maintained due to latent heat H of metal converting from the liquid to the solid state. The evolution of this heat produces the heat flow q_3, Eq. (3.3), which adds to the flow q_1 and keeps the balance:

$$q_1 + q_3 = q_2 \qquad (3.4)$$

Approximately 13 s after immersion (point m), the sample temperature rises to such an extent that the intensity of liquid metal cooling becomes not high enough to

---

[2]In iron-carbon melts, this transition starts at the liquidus temperature and ends at the lower solidus temperature. For the low-carbon steels, as in our case, the temperature difference between these two points can be neglected. Therefore, from now forth we will consider that the metal solidification and melting processes occur at the same constant temperature, i.e. melting point $t_C \cong 1530$ °C.

continue the process of solidification of the layer. The further growth of the layer thickness ceases and the evolving of latent heat H stops. At the point m, $q_3 = 0$ and $q_1 = q_2$. Even greater increase of the sample temperature results in the further reduction of the flow q_2 which becomes smaller than q_1, $q_2 < q_1$. At that, the layer solidification is replaced by its melting. The heat flow q_3 changes its sign. Now this heat is not evolving in the layer, but is consumed by melting of the layer, which ends at the point n. Then the melting of the sample itself starts, which ends at the point x, Fig. 3.2. The same balance equation corresponds to the melting curve in the interval from point m to point x:

$$q_1 = q_3 + q_2 \tag{3.5}$$

At the end of melting, the average-mass temperature of the sample becomes approximately equal to t_C, and the value of q_2 can be disregarded as compared to q_3. At this last interval, the melting curve practically coincides with a straight line. At this, the melting rate, which is characterized by the tangent of the angle γ, increases up to its maximum value, since practically the entire flow q_1 is used for melting of the sample. Extrapolation of the straight line allows determining the position of the point X, and consequently the time of complete melting down of the sample, with sufficiently high accuracy, Fig. 3.2.

Complete elimination or even partial shortening of the period of the solidified layer existence considerably decreases duration of scrap melting. Let us examine the conditions at which scrap can be melted in liquid metal without solidifying. As the analysis of the melting curve demonstrates, solidifying occurs only under conditions of $q_2 > q_1$. Therefore, in order to eliminate solidification, it is necessary to increase the heat flow to the scrap surface q_1 and to decrease the heat flow q_2.

According to Eq. (3.1), the heat flow q_1 rises with an increase of heat transfer coefficient α as well as of temperature of liquid metal t_L. The possibilities of increasing α are discussed below. In regard to the temperature t_L, the formation of the solidified layer is completely eliminated when this temperature is high enough. However, under actual operating conditions of steel production, this temperature is determined by process conditions, and its substantial increase does not seem possible.

The most effective method of elimination of solidification is decreasing the heat flow q_2 by scrap preheating. As analysis of data published in the work [1] demonstrates, the duration of the period of solidified layer existence shortens as preheating temperature t_S increases, and at $t_S \cong 890$ °C or higher solidification does not takes place anymore, Fig. 3.3.

The calculations using the results obtained by the method of sample melting in the works [1–3] have shown that the ratio of duration of existence of solidified layer τ_C to total duration of sample melting τ does not depend, over rather wide range, on the diameter of the sample and is equal to 0.6, Fig. 3.4. This self-similarity of the process is noted here for the first time. It can be used for the analysis of the effect of scrap preheating temperature on duration of melting of not only an individual piece, but also on duration of simultaneous melting of multiple various scrap pieces.

Fig. 3.3 Existing time of
solidified layer, τ_C, versus
sample heating
temperature, t_S

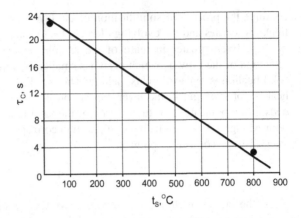

Fig. 3.3 Existing time of solidified layer, τ_C, versus sample heating temperature, t_S

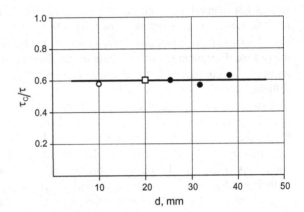

Fig. 3.4 Dependence of ratio τ_C/τ on sample diameter, d. ●—[1]; ○—[2]; □—[2]

3.2.2 Co-melting of Multiple Samples

Co-melting of multiple samples was studied as well at McMaster University, Hamilton, Canada [1]. In these experiments, from two to nineteen cylindrical samples were tied together so that there was a certain gap between them. This gap was varied from test to test. The bundles of tied samples were submerged into liquid metal and then taken out of it at specified time interval. If the gaps were small, the metal between the samples converted into solid state, and metal and samples together formed monolithic conglomerate. The mass of the solidified part of this conglomerate considerably exceeded the overall mass of the solidified layers of metal on samples which were submerged into metal independently of each other. Consequently, the duration of melting of conglomerate increased sharply in comparison with melting of individual samples.

As the gaps between the samples increased, the mass of solidified layers on them decreased, and melting time shortened. When the gaps became quite large, the samples in the bundle melted as fast as they did when they were immersed into metal individually. During these experiments, the relationship between porosity of the bundle of tied samples and its melting time was established. The porosity of the bundle P was determined by the equation:

$$P = 1 - V_{st}/V_{\sum} \tag{3.6}$$

V_{st} volume occupied by samples, m^3
V_{Σ} total volume of the bundle, m^3

As P increased, the duration of melting of the bundles decreased and asymptotically approached the duration of melting of individual samples. At P > 0.96, the bundles melted practically as fast as individual samples [1].

Processing of the results of these experiments allowed to present the relationship under consideration in the form of the most convenient for practical use, Fig. 3.5. The coefficient K_P in Fig. 3.5 shows how many times the melting time τ_{ml} of the bundle of cylindrical samples tied to each other exceeds the τ_{ml} of an individual sample.

The actual conditions for the melting of scrap in a liquid metal are in many respects similar to the melting conditions of the samples in the experiments described above. In shaft furnaces the charging of scrap is always carried out in separate successive portions into quite limited zone of liquid bath. Some pieces of scrap in this zone locate on the bottom and onto each other. At a high rate of charging and a large portion mass, the surface of scrap pieces contacting with metal is significantly smaller than total surface area of individual pieces.

Randomly located in the charging zone, pieces of scrap form a 3-dimensional lattice that has certain hydraulic resistance, which reduces the speed of the metal flows between the pieces. The metal temperature in the zone can drop significantly

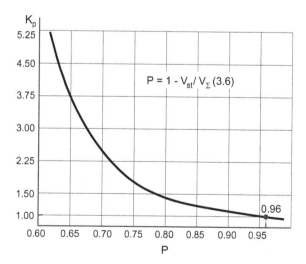

Fig. 3.5 Dependence of K_P coefficient on charging zone porosity P

Table 3.1 Ratio between porosity P in charging zone and bulk density ρ_s, t/m^3

ρ_s, t/m^3	0.50	0.55	0.60	0.65	0.70	0.75	0.80	0.90
P	0.937	0.930	0.924	0.918	0.911	0.905	0.899	0.886

in comparison with the temperature outside of the zone. All these factors decrease heat flow from metal to scrap and slow down the melting. On the other hand, vortices are formed in the flows around the scrap pieces which increase heat transfer intensity in some degree.

By analogy with the melting of the bundles it can be assumed that at P $\geq$ 96% the maximum possible melting rate of the entire mass of scrap located in the charging zone and consequently the maximum possible charging rate are ensured.

3.2.3 Porosity of Charging Zone and Bulk Density of Scrap

It should be emphasized that the porosity P% determines only the share of free space from the scrap in the charging zone. This value is no way connected with the size of pieces of scrap or their quantity in the zone but only with their total mass. The crushing of scrap pieces, provided that they still remain in the charging zone, does not change the porosity of the zone. The granulometric composition of the scrap which largely determines the heat transfer conditions and the mass of the scrap in the zone depend on its bulk density ρ_s, t/m^3. Since the value of P and the relationship given in Fig. 3.5 are further used for calculations of the scrap melting rate, the dependence between P and ρ_s should be established.

Using the equation P = $1-V_{st}/V_\Sigma$, (3.6), and assuming V_Σ = 1 m^3, it is easy to show that there is a unique relationship between P and ρ_s:

$$P = 1 - \rho_s / 7.9 \tag{3.7}$$

ρ_s - bulk density of scrap. t/m^3, 7.9 t/m^3—density of steel.

It is assumed that when the scrap is charged into a liquid metal the value of ρ_s does not change.

The results of calculation according to Eq. (3.7) are shown in Table 3.1.

References

1. Li J, Brooks GA, Provatas N (2004) Phase-field modeling of steel scrap melting in a liquid steel bath. AISTech Conference 1:833–843
2. Fleisher AG, Kuzmin AL (1982) Effect of temperature of the melt on heat transfer to the surface of an immersed melting body. Izwestiya VUZov, Ferrous Metall (4): 40–43
3. Glinkov GM, Bakst VYa, Megzhibozhski MYa et al (1972) Melting a cold steel scrap in the overheated ferricarbonic melt. Izwestiya VUZov, Ferrous Metall 3: 62–64

Chapter 4
Calculations of Scrap Melting Process in Liquid Metal

Abstract To calculate the scrap melting time actual scrap was replaced by its equivalent model. There is no calculation of scrap melting rates (which is of interest in practice) which can be done without such a substitution. The scrap consisting of short steel cylinders with a diameter of 25 mm was adopted as an equivalent model. With regard to the process of heat transfer from liquid metal to scrap, the ratio of the total surface area of scrap pieces F to their overall mass M is the basic characteristic of scrap. This ratio can be assumed as criterion of similarity of the equivalent scrap to the real scrap. The calculations have shown that, despite a considerable spread of data on granulometric composition of scrap used in varied plants, the F/M equal to approximately 20 m^2/t can be assumed as the typical average value for scrap with volume density of 0.6–0.7 t/m^3. Diameter 25 mm corresponds to this ratio. Further adaptation of experimental data to real conditions of scrap melting is carried out by using the system of correction coefficients. On this basis a method for determining the time of melting has been developed. This method can be used as a tool for analyzing the results of the operation of furnaces and when designing them. Examples of calculations of the scrap melting time on the Consteel conveyer furnace and on a shaft furnace of the Quantum-type are given.

Keywords Scrap melting time in liquid metal · Calculation method · Adaptation experimental data to real conditions · Equivalent scrap · Correction coefficients · Scrap charging zone

4.1 Scrap Melting Time

The general expression for the melting time of scrap in case of its continuous charging into the limited zone of liquid metal can be obtained from the heat balance equation of the melting process:

$$\alpha \left(t_L - t_{av}\right) \cdot F \cdot \tau = M[H + c_p(t_{av} - t_s)] \qquad (4.1)$$

© The Author(s) 2017
Y.N. Toulouevski and I.Y. Zinurov, *Fuel Arc Furnace (FAF) for Effective Scrap Melting*, SpringerBriefs in Applied Sciences and Technology,
DOI 10.1007/978-981-10-5885-1_4

M   mass of scrap, kg
F   total surface area of the scrap pieces, m^2
H, c_p latent heat of melting, Wh/kg, and heat capacity of a solid piece, Wh/(kg·°C)
α   coefficient of heat transfer, W/(m^2·°C)
t_L, t_{av}, t_s temperature of liquid metal, average temperature of melting scrap piece,
             and scrap preheating temperature, °C, respectively
τ   melting time, h

The left side of Eq. (4.1) is a quantity of heat obtained by scrap during the time τ, and the right side is heat consumption for melting at a temperature t_{av}. The quantity of heat evolved during solidification of metal layer on scrap pieces is equal to the quantity of heat which is later on consumed for the melting down of the solidified metal. Therefore, neither of these heat quantities is considered in Eq. (4.1).

If α is known, the expression for τ follows from Eq. (4.1):

$$\tau = \frac{M\left[H + c_p(t_{av} - t_s)\right]}{F \cdot \alpha\,(t_L - t_{av})} \qquad (4.2)$$

4.2 Adaptation of Experimental Data Obtained by the Method of Melting Samples to Real Conditions of Scrap Melting

4.2.1 Equivalent Scrap

The above experimental data, Chap. 3, [1], allow calculating the level of an increase in the scrap melting rate in the charging zone which can be achieved by increasing the temperature of scrap preheating and other available means. Such a calculation requires to replace a real scrap with its equivalent model. It should be noted that there are no calculations of scrap melting rates (which are of interest in practice) which can be done without such a substitution.

The scrap comprised of short steel cylinders d in diameter was taken as the equivalent model. This model is most convenient for comparison of the results of calculations with the experimental data obtained by melting of cylindrical samples. With regard to the process of heat transfer from liquid metal to scrap, the ratio of the total surface area of scrap pieces F to their overall mass M is the basic characteristic of scrap. Under otherwise equal conditions, the higher the ratio F/M, m^2/t, the shorter the melting time, Eq. (4.2). This ratio can be assumed as criterion of similarity of the equivalent scrap to the real scrap. The calculations have shown that, despite a considerable spread of data on granulometric composition of scrap used in varied plants, the F/M equal to approximately 20 m^2/t can be assumed as the typical average value for scrap with the volume density of 0.6–0.7 t/m^3. The diameter of 25 mm corresponds to this ratio.

4.2.2 Correction Coefficients K_P, K_L, K_{ts} and K_α

4.2.2.1 Coefficient K_P, Adjustment of Porosity P in the Charging Zone

At values of P greater than 0.96, any amount of the equivalent scrap in the charging zone will be melted down in the same minimum time τ_{min}, as a single piece of equivalent scrap with a diameter of 25 mm. With a reduction in the P caused by an increase in the mass M of scrap in the zone the melting time increases by a factor of K_P in comparison with τ_{min}, Chap. 3, Fig. 3.5.

4.2.2.2 Coefficient K_L, Adjustment of Temperature of Metal t_L

The experiments were carried out in the induction furnace at a temperature of $t_L = 1650\ °C$. In furnaces with a flat bath the minimal permissible value of t_L during the scrap melting period amounts to $1580\ °C$. According to Eq. (4.2), with reduction in t_L from 1650 to 1580 °C the scrap melting time increases by a factor of 2.4, $K_L = 2.4$.

4.2.2.3 Coefficient K_{ts}, Adjustment of Scrap Preheating Temperature t_S

When preheating of scrap to a temperature of t_S the time of scrap melting shortens by a factor of K_{ts}, Fig. 4.1. For example, the melting time at $t_s = 800\ °C$ in

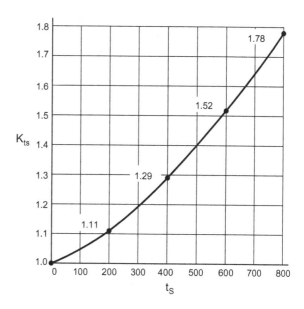

Fig. 4.1 Dependence of coefficient K_{ts} on scrap preheating temperature t_S

comparison with a cold scrap (0 °C) shortens by a factor of 1.78, $K_{ts} = 1.78$. Exactly the same reduction at $t_S = 800$ °C follows from the analytic Eq. (4.2).

4.2.2.4 Coefficient K_α, Adjustment of Metal Stirring Intensity

In baths of EAFs, the liquid metal is intensively stirred by oxygen jets, CO bubbles, and by circulation and pulsation stirring as well [1]. This dramatically increases the heat transfer coefficient α and shortens the melting time, Eq. (4.2). An effect of the metal stirring intensity on α in the experiments with immersed cylindrical samples was studied in the work [2]. In comparison with natural convection in the power-off induction furnace, argon blowing of metal from above increased the coefficient α by a factor of 1.6. The scrap melting time has to proportionally shorten, Eq. (4.2), $K_\alpha = 1.6$.

Using the above correction coefficients the melting time of the mass of equivalent scrap τ_m in the charging zone can be determined:

$$\tau_m = 38 \times K_P \times K_L / K_{ts} \times K_\alpha, \text{ sec} \qquad (4.3)$$

38 s is the melting time of a single cold sample with d = 25 mm in the induction furnace at $t_L = 1650$ °C.

Equation (4.3) relates the time of the melting of equivalent scrap under real conditions of the metallic bath of EAFs with the results of studies of the melting of samples in induction furnaces. The coefficients K_P and K_L increase with decreasing P and t_L. The coefficients K_{ts} and K_α reduce with decreasing the scrap preheating temperature t_s and the intensity of bath stirring. Both of these result in an increase in τ_m.

To increase the melting rate it is necessary to increase the volume of the charging zone, the scrap preheating temperature, and the intensity of bath stirring. As for the temperature of metal t_L which determines the K_L coefficient, it should be maintained during the scrap melting period at the maximum allowable for furnaces with a flat bath level of 1580 °C. At this temperature the coefficient K_L assumes a minimum value of 2.4.

At $K_P = 1.0$ (P >0.96), $K_L = 2.4$ ($t_L = 1580$ °C) and $K_\alpha = 1.6$, the values of τ_m for various scrap preheating temperatures t_s are given in Table 4.1. Let us recall that at $K_P = 1.0$ any amount of scrap in a sufficiently large charging zone will be melted for the same time as a single sample.

It should be emphasized that the melting time τ_m can be determined not only by the conditions of heat transfer in the liquid metal but also by the intensity of the heat input to the metal bath by electric arcs. To maintain the metal temperature at the maximum level of 1580 °C the amount of heat supplied at any time should be equal

Table 4.1 Melting time of equivalent scrap

t_s, °C	0	200	400	600	800
τ_m, s	57	51	44	38	32

to the heat consumption for melting of the scrap. Maintaining the energy balance between the heat input and heat output, at the scrap melting rate corresponding to a temperature of 1580 °C, requires a transformer of a certain power. Other things being equal, but with a lower electrical power the melting rate will drop due to a reduction in the temperature t_L.

4.3   Calculation Method of Scrap Melting Time in Liquid Metal

4.3.1   General Characteristic of the Method

The proposed method, Eq. 4.3, can be used both for analyzing the results of the operation of furnaces and for their design. As already emphasized, the time of melting a scrap in the liquid metal is a key parameter of the entire process and has the great impact on the productivity and electrical energy consumption of the group of furnaces in question.

The proposed method allows quantifying the effect of various design and technological factors on the melting time. It is based on the reliable experimental data obtained in induction furnaces and adapted to the actual melting conditions in the EAF bath. The accuracy of calculations by the proposed method is quite sufficient to use their results in practice. This is confirmed by a basic calculation example given below.

4.3.2   Examples of Calculations of Scrap Melting Time

4.3.2.1   Conveyor Furnace Consteel

The calculation is basic. Its results allow using the proposed method in other conditions, particularly in shaft furnaces with scrap retaining fingers. In Consteel furnaces, all the scrap is almost continuously charged into liquid metal. The bath remains flat. Shaft furnaces with fingers are charged in two-four large portions. In this case, only part of the scrap of each portion is placed in a liquid metal. The rest of the scrap is located above the bath. Under these conditions the use of the method for determining the scrap melting time requires an additional justification which is facilitated by the basic calculation.

A 170-t conveyor furnace of a mini-mill located in Asha, Russia, has been chosen for the basic calculation. Such a choice was due to the fact that this furnace was previously studied in detail. This made it possible to establish with great accuracy all the initial data necessary for calculation and to compare the estimated melting time with actual one.

Initial data: tapping weight is 120 t; hot heel is 50 t; mass of scrap consumed for the heat M_s = 129 t; scrap bulk density ρ_s = 0.65 т/м^3; volume of charging zone V_Σ = 4.3 м^3; K_L = 2.4 (t_L = 1580 °C), K_{ts} = 1.1 (t_s = 200 °C, Fig. 4.1); K_α = 1.6; K_P = 1.1. At ρ_s = 0.65 т/м^3, porosity P of the zone is 0.918, Chap. 3, Table 3.1. K_P = 1.1 corresponds to this porosity, Chap. 3, Fig. 3.5.

The mass of scrap m_s which can be placed in the charging zone amounts to $V_\Sigma \times \rho_s$, m_s = 0.65 × 4.3 = 2.8 tons. Let us determine the melting time of the scrap with a mass of 2.8 t according to Eq. (4.3): τ_{ms} = 38 × 1.1 × 2.4/(1.1 × 1.6) = 57 s or 0.95 min.

The scrap melting rate is 2.8/0.95 = 2.95 t/min. The calculated melting time of the entire scrap amounts to 129/2.95 = 43.7 min. An average actual melting time is about 40 min. In this case the bath remains completely flat on its entire surface including above the charging zone.

Let us determine the minimum transformer power which is necessary to maintain the calculated melting rate of 2.95 t/min without taking into account the possible additional heat losses. In order to melt down 1 ton of scrap and to heat the melt to a temperature of 1580 °C it is necessary to introduce into the liquid metal the amount of heat Q = 379 kWh, Table 1.1, Chap. 1. This total heat input consists of several terms: $Q = q_{el} + q_1 + q_2 + q_3$. According to the calculations the heat input from the oxidation of iron and its impurities is q_1 = 79 kWh/t; the heat input from used fuel (i.e. natural gas, coke powder, lumpy coke etc.) is q_2 = 63 kWh/t; the heat input from scrap preheating to 200 °C is q_3 = 27 kWh/t, Table 1.1, Chap. 1. Now the necessary heat input from the use of electrical energy can be found: q_{el} = 379−79−63 −27 = 210 kWh/t. In the calculations, all the terms of q are determined taking into account the coefficients η which characterize the degree of absorption of heat evolved by the bath (the energy efficiency coefficients). For q_1 and q_2 the coefficient η is equal to 100%. In the calculations of q_2 the values of η are determined for each type of fuel. For q_{el} the coefficient η is equal to 0.84, Chap. 2, Sect. 2.3.2.1, Eq. (2.5).

To melt the entire scrap and heat the melt to 1580 °C it is necessary to feed 210 kWh/t for 43.7 min at η = 0.84. Minimum power of the transformer required for the melting period at $\cos\varphi$ = 0.79 is 210 × 129/(0.73 × 0.84 × 0.79 × 10^3) = 55.9 MVA. The installed power is 90 MVA and cannot limit the melting rate of scrap.

4.3.2.2 Shaft Furnace Quantum

The Quantum furnace of 160-t capacity operating in Mexico is examined.

Initial data: the mass of scrap M_S is 110 t; scrap bulk density ρ_s = 0.65 t/m^3; K_L = 2.4; K_{ts} = 1.29 (t_s = 400 °C, Fig. 4.1); K_α = 1.6; K_P = 1.1; V_Σ = 7.5 m^3. This volume is determined according to Primetals Technologies' data as the product of the area of free section of the shaft at the level of the fingers on the average (during the scrap melting) depth of the metal bath under the shaft.

Scrap is charged into the bath in four portions with an interval of 8.5 min. The mass of each portion m_{por} is 110/4 = 27.5 t. The mass of scrap at the charging zone

$m_s = V_\Sigma \times \rho_s$, $m_s = 4.9$ t. The m_{por} is greater than the m_s by a factor of 5.6. Therefore, after each charging most of the scrap is over the bath. The bath in the area under the shaft becomes flat only for periods of time from immersion of a scrap portion residue into the bath to charging the next batch, see Chap. 3, Fig. 6.3.

The continuous process of settling a portion of scrap and its melting in the liquid metal of the shaft furnace can be calculated in the same manner as the process of continuous charging a scrap into the bath of the conveyor furnace. An additional heating a part of scrap over the bath and in slag is neglected. In this case, the calculated values of τ_{por} will be somewhat overstated.

The scrap melting time of the mass $m_s = 4.9$ t is determined according to Eq. (4.3):$\tau_{ms} = 38 \times 1.1 \times 2.4/(1.29 \times 1.6) = 48.6$ s. or 0.81 min. The melting rate w_s is $4.9/0.81 = 6.0$ t/min. The melting time of the entire portion is $27.5/6.0 = 4.6$ min. The calculated melting time of all the scrap is: $4.6 \times 4 = 18$ min (0.3 h). The actual melting time of one portion is about 7 min and that of all the scrap is $4 \times 7 = 28$ min. Such a significant discrepancy between the calculated and actual melting time is due to insufficient transformer power. With insufficient energy supply a temperature in the charging zone decreases and the melting time considerably increases, Eq. (4.2). In order to maintain the calculated melting rate of the 6 t/min ($t_L = 1580$ °C) the transformer of minimum power rather 110 MVA than 80 MVA is required. This is confirmed by the following calculation below.

In order to melt down 1 t of scrap and to heat the melt to a temperature of 1580 ° C the amount of heat $Q = 379$ kWh/t has to be introduced into the bath. This heat input has to be provided by the four energy sources: electrical energy q_{el}, exothermic reaction of oxidation of iron and its alloys q_{ch}, preheated scrap q_s, carbon-containing materials q_{cok}. As per calculations considering appropriate energy efficiency coefficients we obtain: $q_{ch} = 79$ kWh/t, $q_s = 66$ kWh/t, Table 1.1, Chap. 1, and $q_{cok} = 34$ kWh/t. As the sum of $q_{ch} + q_s + q_{cok}$ is 179 kWh/t then $q_{el} = 379-177 = 200$ kWh/t.

In order to melt down and heat all the scrap in 18 min (0.3 h) it is necessary to input $200 \times 110 \times 10^{-3} = 22.0$ MWh. In this case, with $\eta_{el} = 0.84$ and cos $\varphi = 0.79$, the transformer power of $22.0/(0.3 \times 0.84 \times 0.79) = 110$ MVA is required.

4.3.2.3 Influence of Scrap Quality

An increase in the bulk density of scrap ρ_S significantly increases the melting rate. This can be explained by the fact that when using the higher quality scrap instead of light scrap the mass of the scrap m_s in the charging zone grows faster than the zone porosity decreases and the coefficient K_P increases. For example, at $V_\Sigma = $ constant an increase in ρ_s from 0.55 to 0.7 t/m^3 increases m_s by a factor of 1.27. In this case, the porosity P decreases from 0.93 to 0.91 and K_P increases from 1.0 to 1.06, Chap. 3, Table 3.1, Fig. 3.5. The mass of melting scrap increases by a factor of

1.27, and scrap melting time increases by a factor of 1.06. Therefore, the melting rate increases by $1.27/1.06 = 1.2$ times.

4.3.3 Specific Scrap Melting Rate

Let us introduce a new value namely the specific melting rate of cold scrap ($t_s = 0$ °C) calculated per 1 m^3 of the volume of the charging zone: $w_0 = w_s/V_\Sigma$, $t/(m^3 \cdot min)$. Under identical melting conditions for scrap in a liquid metal, the values of w_0 must also be the same, independent of either the type of a furnace or the volume of the charging zone. This makes the value of w_0 very useful in a comparative analysis of various furnaces. Let us show this on examples of furnaces Consteel, Quantum, and 420-t Tokyo Steel.

When comparing the values of w_0 in the conveyer furnace, Sect. 4.3.2.1, with those in the shaft furnace Quantum, Sect. 4.3.2.2, it is necessary to take into account that in the first case $K_{ts} = 1.1$ ($t_s = 200$ °C) and in the second case $K_{ts} = 1.29$ ($t_s = 400$ °C). Now the values of w_0 in these furnaces can be reduced to the same temperature of $t_s = 0$ °C by means of expression: $w_0 = w_s/(V_\Sigma \times K_{ts})$. In the Consteel furnace the value of $w_0 = 2.95/(4.3 \times 1.1)$; $w_0 = 0.6$ $t/m^3 \cdot min$. In the Quantum furnace the value of w_0 is the same: $6.0/(7.5 \times 1.29) = 0.6$ $t/m^3 \cdot min$. The scrap melting rate is very much dependent on the intensity of stirring the bath when it is blown with oxygen. Therefore, the equality of the obtained values of w_0 means that the stirring intensity of the bath in both cases is also approximately equal.

The Consteel furnace is equipped with ordinary sidewall jet modules which have little effect with regard to bath stirring. In the Quantum furnace, two oxygen lances introduced into the furnace freeboard through the holes in the roof are installed. It is well-known that such lances can provide very high bath stirring intensity. However, this possibility was not realized in this case.

Let us consider the 420-t DC EAF Tokyo Steel. In this furnace scrap is charged into liquid metal by a conveyor as in Consteel furnaces. The tapping weight is 300 tons; the hot heel is 50 tons. The productivity reaches 360 t/hr at a melting rate $w_0 = 9$ m/min.

The data published are insufficient to accurately determine the volume of charging zone. In this furnace, according to approximate calculations the w_0 is 0.9 $t/(m^3 \cdot min)$. Such a high specific scrap melting rate can be explained by the fact that this furnace, along with jet modules, is equipped with an additional oxygen lance which is introduced into the furnace through the slag door and submerged into slag by means of a manipulator. The blowing with immersion of the lances into the slag sharply increases the bath stirring intensity.

The issues of intensification of the scrap melting process in liquid metal by oxygen blowing of the bath are discussed in the next chapter.

References

1. Toulouevski Y, Zinurov I (2013) Innovation in electric arc furnaces: scientific basis for selection. Springer
2. Fleisher AG, Kuzmin AL (1982) Effect of temperature of the melt on heat transfer to the surface of an immersed melting body. Izvestiya VUZov, Ferrous Met Bull (4): 40–43

Chapter 5
Increasing Scrap Melting Rate in Liquid Metal by Means of Oxygen Bath Blowing

Abstract Shaft furnaces with charging a scrap into the liquid metal introduce new fundamentally important features into the known blowing goals. Along with the process of decarburization of the bath the main task of blowing becomes the increase in the scrap melting rate in a limited zone of its charging due to an increase in the metal stirring intensity. The melting rate increases sharply when the lances are immersed into the melt to the slag-metal interface. Thermal work of tuyeres with such immersion is considered. The values of the heat flows to the tip of the tuyere in the slag and the metal are given. The inadmissibility of the cooling mode of tuyeres with local boiling of water and advantages of jet cooling are shown. The design of the tip of a three-nozzle oxygen tuyere using these advantages is described. The tuyere is designed for submerging into the bath to the slag-metal interface near the scrap charging zone. According to the thermal calculation of the tuyere the temperature of the tip surface in the slag does not exceed 72 °C, and in the metal 142 °C. Such low temperatures even in case of submerging in metal make it possible to expect the very high durability of the tuyeres with jet cooling. The automated control system for the position of the tuyeres is necessary to realize efficient oxygen blowing of the bath at the slag-metal interface in conditions of short tap-to-tap times when the liquid metal surface level raises rapidly. The system must continuously receive information from the sensor fixing the position of the tuyere tip relative to the metal level. A reliable sensor tested in industrial conditions is suggested.

Keywords Intensity of metal stirring · Oxygen tuyeres · Thermal operation of tuyeres · Immersion of tuyeres into melt · Slag-metal interface

5.1 Preliminaries

Shaft furnaces with charging a scrap into the liquid metal introduce new fundamentally important features into the known blowing goals. Along with the process of decarburization of the bath the main task of blowing becomes the increase in the scrap melting rate in a limited zone of its charging due to an increase in the metal

© The Author(s) 2017
Y.N. Toulouevski and I.Y. Zinurov, *Fuel Arc Furnace (FAF) for Effective Scrap Melting*, SpringerBriefs in Applied Sciences and Technology, DOI 10.1007/978-981-10-5885-1_5

61

stirring intensity. In shaft furnaces the charging zone of scrap and the zone of heating of liquid metal by electric arcs are at a certain distance from each other, Chap. 2, Fig. 2.9. At an increase in the scrap charging rate it is not enough to increase the arc power in order to avoid a local drop in the temperature of the metal in the charging zone and a corresponding increase in the scrap melting time. It is also necessary to increase the intensity of stirring the metal between these zones. Jet modules overall wide-spread in EAFs are not very suitable for solving this task, Chap. 2, Sect. 2.2.2. The distance between oxygen nozzles of the module and the surface of the liquid metal is too great, especially at the beginning of the melting period.

In order to increase the efficiency of modules when blowing the bath the oxygen jet is surrounded with an annular cocurrent flow of the pilot flame of the burner. The gases in the flame have low density due to the high temperature. In this case, the known effect of increase in the range of the jet flowing into the less dense ambient medium is used. However, let us also take into consideration a fact that, even without the surrounding pilot flame, the oxygen jets propagate in the EAF freeboard filled with gases with the temperature in the order of 1600 °C. These gases contain a large amount of CO. Taking into account post-combustion of CO in the O_2 jets, the density of the gases is not much different from the gas density in the pilot flame. Therefore, the additional increase in the long range of oxygen jets resulting from surrounding the jets with the pilot flame cannot be considerable.

The general use of the pilot flame in the modules, which requires significant additional consumption of natural gas, is required not so much for increasing the range of oxygen jets, as for its ability to facilitate penetration of oxygen jet to the melt through the layer of scrap. At the same time, the probability of reflection of oxygen jets from the scrap lumps to the water-cooled elements of the modules decreases. However, such an effective function of the pilot flame is excluded in the furnaces with flat bath since scrap is absent in their freeboard. In respect of increasing the long range of oxygen jets, there are no reliable data obtained under industrial conditions, which confirm an improvement in effectiveness of oxygen bath blowing, due to that increasing. At the conveyor furnace of 170 t capacity, Asha, Russia, the abandonment to use the pilot flame, when blowing the bath with jet modules, allowed reducing natural gas flow rate and had no effect on the furnace operation performances.

The second method of increasing the jets long range consists in configuring the optimum profile of the supersonic nozzle, which minimizes the turbulence of the flow. By analogy with the lasers, these nozzles and the supersonic jets formed by them are called coherent. Due to the quite low turbulence, the coherent jets involve into their motion considerably smaller mass of the ambient gas, expand slower, and maintain initial velocity at the considerably greater distance from the nozzle, in comparison with the jets flowing from the simple de Laval nozzles.

In the tests conducted in the laboratories on the testing ground, the achieved increase in the length of the initial region of the coherent jets was by approximately 5 times more in comparison with that of the jets flowing from the simple de Laval nozzles.[1] Such results of the stand tests can make a false impression regarding the very broad possibilities of coherent nozzles. However, it is necessary to understand that the conditions of the coherent nozzles operation in the EAF differ considerably from the laboratory conditions. Even the simple de Laval nozzles are quite sensitive to the most insignificant deviations of the oxygen parameters from the design parameters. To an even greater degree, this relates to the coherent nozzles. It is noted that small deviations from the design operating conditions eliminate the advantages of these nozzles, whereas under the production conditions, such deviations occur constantly. Furthermore, even quite small deposits of droplets of metal and slag inside the nozzles, which cannot be avoided in practice, cause strong turbulization of the flow disrupting its coherence. It is also necessary to note that there are no any direct or indirect data obtained under the actual conditions in the furnaces, which could confirm the results of the stand tests. All the aforesaid leads to the conclusion that the known methods of increasing the long range of oxygen jets cannot compensate a large distance of the jet modules from the surface of the metal bath. It is impossible to sharply increase the intensity of bath stirring in the charging zone of scrap into the liquid metal by means of jet modules. For this purpose some different more effective devices have to be used.

5.2 Tuyeres with Evaporation Cooling Embedded in the Lining

It is necessary to minimize the distance between tuyere nozzles and a liquid metal surface so that oxygen jets could intensively stir the metal bath. Maximum effect is ensured by the immersion of oxygen tuyeres into the melt down to slag-metal interface. This is confirmed by both the results of simulation and long-term international experience of blowing a bath of open-hearth furnaces. In open-hearth furnaces, the submerged blowing has overall replaced blowing with tuyeres positioned above the bath as soon as roof tuyeres which possess high enough durability were developed. It should be emphasized, that hydro-dynamical processes in the bathes of hearth furnaces such as open-hearth and electric arc furnaces are completely similar. Therefore, the open-hearth experience of the bath blowing is of a great interest for modern EAFs as well.

In the second-half of the nineties, understanding the fundamental advantages of the submerged blowing resulted in the development by the KT-Köster Company, of the blowing tuyeres of a new type, the so-called KT-tuyeres. These tuyeres are installed in the lining of the banks of bottom of the furnace. According to the initial concept, the oxygen KT-tuyeres had to be installed slightly lower than the slag surface, and the KT-tuyere for the carbon injection even lower, i.e. near the slag-metal interface. The

---

[1]Let us recall that initial region is the region of the jet where its axial velocity does not decrease and remains equal to the initial velocity.

water for KT-tuyere cooling is used in the mixture with the compressed air. The water is atomized in the heat-stressed frontal part of the tuyere near the head. The small drops of water evaporate on the extended surface of the head. There is no water in the tuyere head itself. In case of the tuyere burn-back, only a small amount of highly atomized water can get into the liquid metal which is no danger.

In the KT system, the heat is mainly removed rather due to evaporation of water than due to heating of it as in the usual water-cooled devices. It is known that the quantity of heat required to heat water from 0 °C to the boiling point of 100 °C is approximately 5 times less in comparison with the quantity of heat required for its total evaporation at this temperature. This allows reducing water consumption for cooling of KT-tuyeres multi-fold as compared to that for cooling of jet modules. The mixture of air, residual portion of atomized water, and water vapor is sucked off from the tuyere by the vacuum pump. As a result, there is no excess pressure in the tuyere. After exiting the tuyere, the air separates from the water and exits into the atmosphere, the vapor condenses, and water is returned into the cooling system of the furnace.

Installation of stationary water-cooled tuyeres and other water-cooled elements in the lining of EAFs' bottom banks even higher than the level of the sill of slag doors is with good reason considered as highly dangerous. The accumulated practical experience shows that, in this case, visible burn-backs of the tuyeres do not pose a special hazard. The visible burn-backs are rapidly detected due to a number of signs indicating the burn-backs, which allows to promptly stop the water leakage by switching off the burnt tuyere. Small hidden leakages, which are difficult to detect on time, pose a real danger. Such leaks damage the internal, hidden from observation lining layers resulting in severe accidents with explosions and metal escape. Therefore, the installation of water-cooled tuyeres in the bottom banks lining must be used only with the application of highly reliable technical equipment, which either guarantees practically inertialess detection of any smallest hidden water leakages or eliminates any possibility of such leakages. The KT-system completely satisfies this requirement especially due to a negative pressure inside the tuyere. A complete safety of installation in the bottom banks lining can be ensured by the tuyeres cooled by high-pressure water circulating in a short closed circuit as well [1]. However, such very promising systems, unlike the KT-tuyeres, have not yet found a practical application in EAFs.

The industrial trials of the KT-tuyeres have shown that the cooling system utilized in them does not allow the contact of the tuyere head with the liquid metal. Therefore, later on, not only oxygen KT-tuyeres have been installed only into the slag close to its surface at the sufficiently great distance from the slag-metal interface, but carbon KT-tuyeres as well. As the slag foaming takes place, the heads of the KT-tuyeres are submerged into the slag to a significant depth. It is impossible to install the blowing devices of the jet module type at that height.

Compressed air is used for injecting of carbon powder into the slag. Natural gas, which is introduced through the annular gap surrounding the oxygen jet, is delivered to the oxygen KT-tuyeres. The flow of gas protects the refractory from the contact with oxygen and, thus, prevents rapid wear of the lining. Compressed air and oxygen provide protection from the slag flowing in the KT-tuyeres. Similar to jet modules, oxygen KT-tuyeres are used for blowing of the bath during the liquid bath stage, and used as burners for scrap heating at the beginning of the heat. Oxygen and carbon KT-tuyeres are installed in pairs, next to each other. These pairs are dispersed along the entire perimeter of the bath. The oxygen, carbon, natural gas and compressed air feeding to the tuyeres is fully automated.

KT-tuyeres have found a certain use. The obtained results confirm a sharp increase in the effectiveness of blowing when the tuyeres are submerged into the slag. In this respect, the most demonstrative are the data obtained from 100-t DC EAF operating on a charge material containing from 80 to 90% of metalized pellets. The pellets are charged into the bath by a conveyor through the opening in the furnace roof. Before replacement of manipulators with consumable pipes with the KT system it took 60 min to charge 65 t of pellets. It was impossible to charge the pellets faster, because this led to the formation of large "icebergs" from the unmolten pellets floating on the bath surface. When working with the KT system, 74 tons of pellets were charged in 33 min. In this case, the furnace productivity with regard to the pellets melting process increased more than doubled. These results could be obtained only in case of radical increase in the intensity of bath stirring by oxygen jets due to submerging of the heads of the KT-tuyeres into the slag. The same can be said about the results obtained from the 100-t EAF of another plant operating on scrap. Installation of the KT-tuyeres has led to a sharp acceleration of melting of the large lumps of scrap submerged in the liquid metal. This has allowed to charge into the furnace considerably larger lumps of scrap and, thus, to decrease the cost of scrap preparation for melting without decreasing the productivity. The necessity of submerged blowing when melting a scrap in the liquid metal is convincingly confirmed by the results obtained from KT-tuyeres applications.

In addition to advantages resulted from the stirring intensification, the submerged blowing of the bath by the KT-tuyeres has increased the durability of the central refractory part of the furnace roof by 2-3 times, consequently, the consumption of refractory materials and repair costs have been reduced. This is explained by the fact that the durability of the roof refractory is directly related to intensity of bath splashing which is sharply decreased when submerging the tuyeres into the slag. Such a relationship is persuasively confirmed by the results of the studies of splashing conducted by the methods of physical modeling [2]. However, in case of the KT-tuyeres installed in the slag, the evaporative cooling does not provide their sufficiently high durability. Although the design of KT-tuyeres allows for the possibility of replacing the worn part of the head without dismantling of the entire tuyere, their maintenance costs, in comparison with the jet modules proved to be quite significant. This has largely limited their wide use. It is not possible at all to use KT lances for the most effective blowing option at the slag-metal interface since the heads of these tuyeres cannot withstand even a brief contact with the liquid metal.

5.3 Roof Water-Cooled Tuyeres for Bath Blowing at Slag-Metal Interface

5.3.1 Thermal Operation of Tuyeres: Heat Flows, Temperatures

Being submerged into a melt, a tip of the tuyere and a part of its side surface are affected by the heat flows of very high density. If the necessary conditions of heat transfer are not observed, these heat flows, when passing through the tuyere surface to water, can cause impermissible overheating of the external wall of the tuyere, its rapid wear and burnout. Providing the necessary durability of tuyeres is of paramount importance for the implementation of submerged blowing. Under stationary conditions, when the total amount of heat obtained by the tuyere is consumed for heating of water, the thermal operation of the submerged part of the tuyere is determined by the following dependences: the heat balance Eq. (5.1) and the heat transfer Eqs. (5.2), (5.3), (5.4), Fig. 5.1.

$$Q = q_L \cdot F_1 = q_w \cdot F_2 \tag{5.1}$$

$$q_L = \alpha_L(t_L - t_1) \tag{5.2}$$

$$q_L = \lambda/\delta(t_1 - t_2) \tag{5.3}$$

$$q_w = \alpha_w(t_2 - t_w) \tag{5.4}$$

$Q; q_L$ heat flow, W, from melt to tuyere surface, and density of this heat flow, W/m^2

F_1 surface area of submerged part of the tuyere, m^2

q_w density of heat flow from wall to water, W/m^2

F_2 – area of the wall contacting water, m^2

α_L – coefficient of heat transfer from melt to tuyere surface, W/(m^2 °C)

t_L temperature of the melt, °C

t_1, t_2, t_w temperatures, °C, of tuyere surface, of surface contacting with water, and of water, respectively

q_λ density of heat flow in the wall, W/m^2

λ coefficient of thermal conductivity of the wall, W/(m^2 °C)

δ wall thickness, m

α_w coefficient of heat transfer from wall to water, W/(m^2 °C)

Equation (5.1) shows that the process under consideration is a stationary process in which the amount of heat accumulated by the tuyere does not vary. Under the stationary conditions, the entire heat flow obtained by the tuyere surface from the melt (amount of heat per time unit) is transferred to water. It follows from Eq. (5.2) that the heat flow from the melt per unit of the tuyere surface (density of the heat flow) is determined by the external factors and does not depend on the parameters

Fig. 5.1 Single-nozzle tuyere. *1*—bath surface; F_1 and F_2—areas of surfaces shown by heavy lines (designations are given in the text)

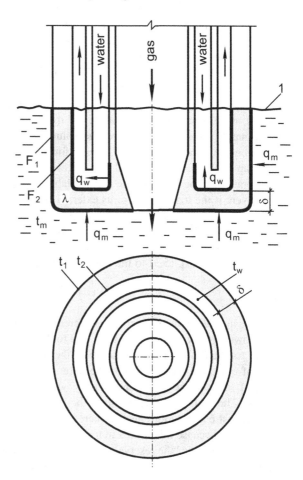

of the tuyere itself. Equation (5.3) explains the heat transfer through the tuyere wall when the wall thickness δ is small and its curvature can be neglected. In actual practice, the tuyeres for bath blowing always satisfy this condition. Finally, Eq. (5.4) determines the tueyre parameters which affect the intensity of heat transfer to water.

Typically, the heat flow density q_L is distributed over the tuyere surface non-uniformly. Usually, the maximum values of q_L are observed near the oxygen nozzles and in the places exposed to blow-backs of oxygen jets reflected from the unmolten scrap pieces. When the tuyeres submerging into the melt come into contact with the scrap, it often causes their burnout. In shaft furnaces, such burnouts are excluded because tuyeres are immersed into slag outside the scrap charging zone.

Under the actual conditions of the bath blowing by the submerged tuyeres, it does not seem possible to determine the values of q_L analytically with acceptable accuracy. Therefore, the experimental data should be used in calculations.

Unfortunately, measurements of q_L under operating conditions of submerged blowing were conducted only in the open-hearth furnaces. The most reliable data are given in the publication [2].

The obtained values of q_L are 1300–1500 W/m^2 in the slag and 4000–5000 W/m^2 in the metal. Triple or even quadruple increase of q_L in the metal is mainly explained by the respective increase of α. In the open-hearth furnaces and in the shaft arc furnaces, the temperatures of the bath during scrap meting stage are approximately the same. Therefore, the value of q_L given above can be used for calculating the parameters of the tuyeres for submerged blowing in the furnaces with scrap melting in liquid metal. Let us review the temperature characteristics of heat operation of the tuyeres. The most reliable indicator of the potential durability of the tuyere is the temperature of its external surface t_1, Fig. 5.1. Rather prolonged period of service is possible if the values of t_1 are low. And the rapid wear of tuyere usually occurs in case of high temperature of its surface. Let us analyse the factors affecting this parameter.

Let us convert Eq. (5.3) to the equation of the form:

$$t_1 = t_2 + q_L \cdot \delta/\lambda \qquad (5.5)$$

The value $q_L \cdot \delta/\lambda$ is the temperature differential in the tuyere wall $\Delta t = t_1 - t_2$. This value grows directly proportional to q_L and to the thermal resistance of the wall δ/λ. In order to obtain the low temperature of the tuyere surface t_1 in case of the high values of q_L, the external wall of the submerging part of the tuyere is made of copper of the highest purity and density which has the maximum thermal conductivity. The wall thickness δ should be minimum allowed by condition of satisfying the strength requirements. Increasing the wall thickness as a reserve is unreasonable, because this results in the increase of t_1 and, respectively, of the wear rate, which surpasses "the reserve". High durability of the tuyeres is achieved when temperatures of the outer surface of the copper wall are lower than 200 °C. The wear rate sharply increases at $t_1 > 300$ °C.

Ensuring low enough temperature level of the inner surface of the wall t_2 is especially important for durability of the tuyeres. This temperature is not just one of the addends determining the value t_1, Eq. (5.5). It also affects the mechanism of heat transfer from the wall to water. Variations in this process can lead to sharp acceleration of wear and rapid burnout of the wall.

The temperature t_2 is defined by the expression (5.6) which follows from Eq. (5.4):

$$t_2 = t_w + q_w/\alpha_w \qquad (5.6)$$

Equation (5.6) shows that the temperature t_2 depends not only on the water temperature t_w and the coefficient of convective heat-transfer α_w, but on the density of the heat flow from the inner surface of the wall to water q_w as well. Despite the fact that, in case of stationary mode, the same heat flow Q invariably passes through the external and the internal surfaces of the wall of the submerged part of the tuyere, Eq. (5.1), the densities of the flows on these surfaces q_L and q_w may differ significantly depending on the ratio between the areas of the surfaces F_1 and F_2, Fig. 5.1. In the past, resulting from then existing technology of the tuyeres manufacturing, the design of the tuyeres for the open-hearth furnaces was such that the area of the surface F_1 considerably exceeded the area of the surface F_2. In these tuyeres, the concentration of the heat flows on the water-cooled surface occurred, which resulted in the increase of the q_w in comparison with q_L by 1.5–2.0 times, which led to the maximum values of q_w equal to about 3×10^3 W/m^2 in the slag and 10×10^3 W/m^2 in the metal.

In many instances, the heat flows of this high density could not be transferred to water by the regular convective heat transfer from the inner surface of the wall. At this surface, the local water boiling occurred in the boundary layer. The very mechanism of the heat transfer changed, which adversely affected the durability of these tuyeres.

5.3.1.1   Operation of Tuyeres with Local Water Boiling

With increasing q_w, when α_w и t_w are constant, the temperature t_2 increases linearly and the lower the coefficient α_w, the quicker this increase is, Eq. (5.6). However, such an increase is observed only to a defined limit. As soon as t_2 exceeds the water boiling temperature t_b at the given pressure, Table 5.1, by a few degrees, the further increase in t_2 slows down sharply. This can be explained by occurrence and development of the local bubble boiling of water which significantly intensifies the heat transfer from the wall.

The mechanism of this process is as follows. When $t_2 > t_b$, despite the fact that water flow in general remains cold, the vapor bubbles are formed on the wall. When they detach from the wall, they pass through the boundary layer of water and condense when entering the cold core of the flow. These bubbles turbulize the laminar sub-layer of the boundary layer; its thermal resistance to the heat flux drops sharply, and the coefficient α_w increases respectively. As q_w grows further, the number of bubbles formed per unit time grows, and α_w continues to increase.

However, increase of α_w continues only to a certain heat flux level critical for the given conditions, $q_{w.cr}$. When the heat flux exceeds this level, the number of bubbles becomes so huge that they coalesce into a solid vapor film which has negligible thermal conductivity and completely insulates the wall from water. If the vapor film is stable, this leads to the abrupt growth of t_2 and t_1 to such high

Table 5.1 Water boiling temperature t_b depending on pressure p, abs

p, bar	1	2	3	4	5	6	7	8	9	10
t_S, °C	99.5	120.4	133.7	143.7	151.9	158.9	165.1	170.5	175.4	179.9

temperatures that heat transfer occurs through radiation only. In this case, the wall burns through immediately.

When the local water boiling is absent, a coefficient of convective heat transfer α_w in Eq. (5.6) depends mostly on the water velocity w. It can be calculated by the simplified Eq. (5.7) [2]:

$$\alpha_w = A \cdot \frac{w^{0.8}}{d_h^{0.2}}, \quad kW/(m^2\,°C) \tag{5.7}$$

The coefficient A is determined by physical properties of water which depend on temperature. Its values are cited in Table 5.2.

d_h—hydraulic diameter of a channel, m.

For the round channels, $d_h = d$. For the channels of rectangle shape, $d_h = 4F/2$ $(a + b)$, where F is cross-section area of the channel, a is the channel width, b is its height, and $2(a + b)$ is perimeter of the channel. For the relatively narrow ring and rectangular channels (slots) with the width of a, $d_h = 2a$.

Since the coefficient of heat transfer to water in the local boiling mode increases, it can seem that the cooling tuyeres in this mode can be effectively used for removal of large heat fluxes. However, in reality, this mode has significant drawbacks and cannot be recommended. Firstly, when the local boiling occurs, the temperature t_2 cannot be below water boiling point t_b which exceeds 140–150 °C at the excessive water pressure in the element of 3–4 bars, Table 5.1. The temperature of the outer surface of tuyeres t_1 increases accordingly, which enhances wear.

Secondly, the local boiling mode is potentially dangerous due to the fact that under certain unfavorable conditions the bubble boiling in some local areas of the element can change to the film boiling which would immediately lead to burn-back. One should not count on the fact that when water velocity is quite high, about 3–5 m/s, the steady film boiling requires such high heat fluxes which are not achievable in the practical operation of EAFs. The research carried out on the transparent models of tuyeres with the aid of high-speed filming has shown that when configuration of the channels is complex, the high-frequency pulsations of water velocity occur in these channels. The water velocity can sometimes for the short periods of time drop to very low values in the poorly flown-around sections of the channels in turbulent zones behind the sharp corners, where the water flow detaches from the walls. In such zones the bubble boiling and the film boiling can continuously take turns. This does not lead to an immediate burn-back, but sharply increases the temperatures t_2 and t_1 and the wear rate. The specifics discussed above lead to the conclusion that in order to assure reliability and durability of the water-cooled tuyeres, the modes of operation with local boiling should be avoided.

Table 5.2 The dependence of the A- coefficient at Eq. (5.7) on the water temperature

t, °C	20	25	30	35	40	45	50	55	60
A	2.56	2.59	2.63	2.67	2.71	2.74	2.78	2.80	2.83

The cooling system should be designed in such a way that the temperature of the wall surface flown around by water t_2 is significantly lower that the boiling point t_b, Table 5.1.

5.3.1.2 Jet Cooling

The density of the heat flow q_L from the melt to the submerged surface of the tuyere is distributed quite unevenly over this surface. A part of this surface where q_L reaches its maximum values is a critical zone of the tuyere. It is obvious that to avoid rapid wear and sharp shortening of the service life of the tuyere, its critical zone must be cooled most intensively. In the furnaces with flat bath, the tuyere cannot be submerged into the melt in the scrap charging zone. Therefore, the danger of the adverse effect of the oxygen jets reflected from the scrap pieces on the tuyere is eliminated. In this case, a part of the surface of the copper tip of the tuyere close to the oxygen nozzles is usually a critical zone.

To be efficient, the cooling system of the tuyere must be calculated so as to eliminate the possibility of quite undesirable local boiling of water in the boundary layer. For this reason, the cooling mode of the critical zone in case of local boiling is not further examined. In the absence of local boiling, the cooling intensity is defined by the coefficient of convective heat transfer α_w. In order to increase this coefficient, it is necessary to increase the speed of water w, Eq. (5.7). With that, the problem of the required water pressure appears, since, with increase of w, both the hydraulic resistance and corresponding required water pressure increase directly proportional to w^2.

The hydraulic resistance is equal to a difference in the water pressure at the entrance and at the exit of the tuyere: $\Delta P_w = P_1 - P_2$. In furnace operation, this pressure difference is always invariably limited. In order to supply water to mobile tuyeres, the supply and drain collectors are used in most cases. These collectors are present in every furnace and supply water to the wall and roof panels as well as to the other water-cooled elements. However, the pressure difference between these collectors usually does not exceed 2.5 bars.

So far, installing the special pumps for supplying high pressure water to mobile tuyeres was avoided, since it involves additional costs growing rapidly with an increase of pressure. Hence, the designers of the mobile tuyeres for submerged blowing encounter the problem of ensuring reliable cooling of the critical zone with the minimal hydraulic resistance of the entire tuyere. In this problem, the minimum required water flow rate per tuyere V_w is a preset initial value which can be calculated using Eq. (5.8) with the error not exceeding 1%:

$$V_w = Q/1.15 \cdot \Delta t_w, \quad m^3/h \qquad (5.8)$$

Q maximum heat flow absorbed by water, kW
Δt_w allowable increase of water temperature in the tuyere, °C

When chemically unprocessed water is used, the value Δt_w should be selected so that the final water temperature at the tuyere exit does not exceed 45 °C to avoid salts precipitation. In any case, increasing Δt_w raises the temperatures t_1 and t_2 and increases probability of local boiling occurrence. That is why it is usually assumed in the calculations that the difference Δt_w should not exceed 30 °C.

Equation (5.7) relates to the flow of water along the cooled walls, for example, to the flows in the pipes. In this case, α_w increases mainly due to the decrease of the thickness of the laminar sub-layer which is caused by the increasing in value transverse turbulent fluctuations of the velocity of the flow. Another more effective method of increasing α is known, when high-speed jets of the cooling water are directed onto the wall at a right angle. This method is widely used in practice.

The jets leaving the openings at high speed w_0 hit the cooled surface and spread over it in all directions. In the impact zone, the boundary layer and its laminar sub-layer are being completely destroyed, which results in very high values of α, which increase with an increase in the speed w_0. The coefficient α drops rapidly with distance from the impact zone. That is why the average values of α W/(m^2 °C) strongly depend on the pitch between the jets in both the transversal and axial directions, Fig. 5.2.

The possibility to alter the intensity of cooling of different sections of the water-cooled element by varying the pitch between the jets x is the fundamental advantage of jet cooling. This allows to concentrate in the critical zone of the tuyere the largest number of the jets per unit area of the cooled surface and to gradually

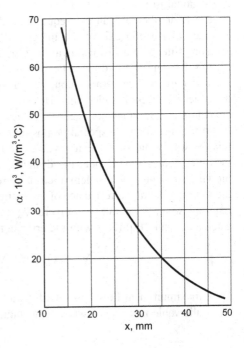

Fig. 5.2 Heat transfer coefficient α versus the pitch between jets, x

reduce the intensity of cooling, increasing pitch x in proportion to distance from the critical zone. None of the other cooling methods has these options.

As it was already noted, in order to achieve maximum values of α, it is necessary to use the maximum possible share of the available water pressure difference for attaining the maximum jet speeds w_0 in the critical zone. For this purpose, the hydraulic resistance of all other sections of the water circuit of the tuyere must be reduced to a possible minimum. In case of given water flow rates V_w and w_0, the number of the jets n increases sharply with the decrease in the diameter of openings d. Directly proportional to the increase in n per unit area of the cooled surface, the pitch between the jets decreases, which, as shown in Fig. 5.2, is accompanied by an increase in α to the values which are practically unattainable in case of the longitudinal flow of water along the cooled surfaces, because required values of velocity and hydraulic resistance are too high. This allows ensure the high durability of the tuyeres with the jet cooling even under the severe conditions of the submerged blowing in case of the extreme values of the heat flows.

5.3.2 Roof Tuyere with Jet Cooling; Design, Basic Parameters

In order to considerably accelerate scrap melting with the help of the tuyeres, they must be submerged into bath close to the charging zone, in much the same manner as shown in Fig. 5.3. The oxygen jets of the tuyeres directed towards the charging zone create the co-current flows of the melt, which increase the intensity of stirring the metal between the charging zone and the zone of arcs.

In Fig. 5.4, a schematic diagram of the design of the tip of a three-nozzle oxygen tuyere using the advantages of jet cooling is shown. The tuyere is designed for submerging into the bath to the slag-metal interface near the scrap charging zone. The tip has a spherical shape which is characterized by the minimum surface area. This is very important for the decrease in the required number of the cooling jets. The tip has water supply chamber (1) and the jet cooling chamber (2). These chambers are separated by the wall (3) with openings (4) forming the water jets which are directed towards the spherical wall of the tuyere (5) at the 90° angle. The closer to the critical zone, i.e. the section of the surface with the nozzles, the smaller the pitch between the openings.

At the exit from the jet cooling chamber, the partition wall (3) narrows down the external ring-shaped channel (6) through which water is removed from the tuyere. This is necessary for increasing the velocity of water and the intensity of cooling of the part of the external pipe which can be submerged in the foamed slag. Above, outside the slag zone, where the density of the heat flows affecting the tuyere sharply decreases, the ring-shaped channel (6) is wider, which reduces the velocity of water through the entire remaining length of the tuyere as well as its general hydraulic resistance. The oxygen is delivered to the tuyeres through the central pipe (7).

Fig. 5.3 Positioning roof tuyere (*1*) in comparison with jet module (*2*)

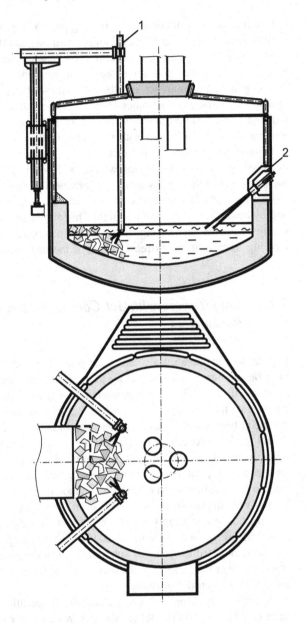

Three lateral supersonic oxygen nozzles are tilted at different angles with regards to the vertical. The tuyere is positioned in the holder arm in such a way that all the nozzles are directed toward the scrap charging zone.

The device, shown in Fig. 5.4, satisfies all the stated above basic principles of designing of the highly durable oxygen tuyeres. The drop of water pressure in the jet cooling zone is 87% of the total hydraulic resistance of the tuyere. The

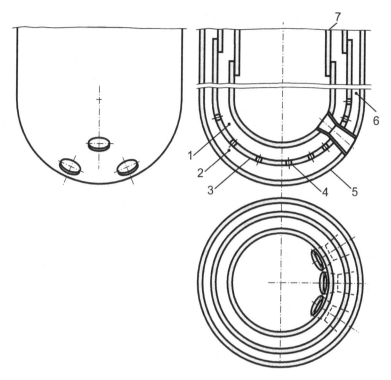

Fig. 5.4 Tip of oxygen roof tuyere (designations are given in the text)

concentration of heat flow on the water-cooled surface is so insignificant that it can be ignored. By reducing the pitch between jets to 15 mm the heat transfer coefficient to water α_w in the danger zone of the oxygen nozzles reaches 65 kW/(m² ·°C) which is 2–3 times higher than the regular level.

Calculations of wall temperatures (5) using Eqs. (5.1)–(5.4) lead to the following results. The external surface of the wall (5) has a temperature of t_1, and a temperature of internal surface flown by water is t_2. Since the thickness of the wall is small as compared with its radius the densities of the heat flow on both surfaces can assume to be identical. When the tuyeres are submerged into the slag the q_L is equal to 1400 kW/m². A temperature t_2 is determined by Eq. (5.6): $t_2 = t_w + q_w/\alpha_w$ which follows from Eq. (5.4). Let us assume that a water temperature at the jet cooling area t_w is equal to 20 °C because it does not practically differ from a temperature at the tuyere inlet. At the $t_w = 20$ °C, $q_L = 1400$ кВт/м², and $\alpha_w = 65$ kW/(m² ·°C), the temperature t_2 will be equal to: 20 + 1400/65 = 41 °C.

When immersing the tuyere tip into liquid metal the q_L will be equal to 4500 kWt/m² and, consequently, $t_2 = 20 + 4500/65.2 = 89$ °C. Thus, even in the case of contact with the liquid metal an occurrence of local boiling in the danger zone of the tip is eliminated.

A temperature of the surface of the tuyere tip t_1 is determined by Eq. (5.5): $t_1 = t_2 + q_L \, \delta/\lambda$, which follows from Eq. (5.3) and where $\delta = 0.0085$ is a thickness of the copper wall, m; λ—the coefficient of thermal conductivity of copper depending on an average temperature of the wall. The value of the λ is resulted from the preliminary calculations of the t_1. The temperature $t_1 = 41 + 1400 \times 0.0085/0.39 = 72 \, °C$ when the tip is in the slag and $t_1 = 41 + 4500 \times 0.0085/0.38 = 142 \, °C$ in the liquid metal, respectively. Such a low temperature of the surface of the most heat-stressed section of the tip even in case of submerging in metal makes it possible to rely on the very high resistance of the tuyeres with jet cooling.

5.3.2.1 Controlling the Optimal Position of Roof Tuyere Relatively to Slag-Metal Interface

The automated control system for the position of the tuyeres is necessary to realize efficient oxygen blowing of the bath at the slag-metal interface in conditions of short tap-to-tap times when the liquid metal surface level raises rapidly. The system must continuously receive information from the sensor fixing the position of the tuyere tip relative to the metal level.

As a result of the persistent attempts to solve this problem for the open-hearth furnaces, the sensors using different principles of operation have been tested [2]. All of them can be divided into two groups. The first group are the sensors built into tuyere. These include pneumatic, radiation, temperature, vibration, and other sensors. All of them proved to be insufficiently reliable. Furthermore, they all make the tuyere design and its manufacturing technology more complicated. For these reasons the sensors built-in the tuyere did not find a practical use.

The sensors of the second group are installed outside of the tuyere. They include the sensors which detect the tuyere weight loss, change in the cooling water temperature, change in the electromotive force generated in the tuyere or electrical resistance between the electrically isolated tuyere and the furnace shell. Out of all of these, only resistance sensors have accuracy and reliability required for the use in the furnaces.

A typical change in the electrical resistance of tuyere in relation to its position in the furnace is shown in Fig. 5.5, [2]. When the tip of tuyere is positioned near the furnace roof, the resistance of the tuyere is 25–30 kOhms. As the tuyere is lowered down, the resistance drops sharply reaching the value of several ohms at the moment when it contacts the slag. In case of further submerging into the slag, the resistance decreases considerably slower, and remains practically constant after the tuyere reaches the metal. Inflection points a and b on the graph correspond to the position of the tuyere on the phase interfaces: gas—slag and slag—metal, Fig. 5.5.

It should be noted, that the absolute value of resistance does not allow determining the position of the tuyere in the phases of the freeboard. The conductivity of the gas phase is determined by its ionization which depends on the temperature and

Fig. 5.5 Electrical tuyere
resistance R versus its
position. Range of pulsations
is shown by *dashed lines.*
Points "a" and "b" are
interfaces of phases

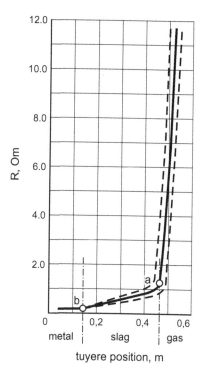

a number of other factors. The same relates to the slag layer located between the tip of the tuyere and the metal. Its resistance depends not only on the thickness of this layer, but also on the temperature, composition and consistency of slag which are changing in the course of the heat.

After the tip gets in contact with the metal, the current passes through the circuit consisting of the metal and inner layer of lining. Because of the large cross section and low specific resistance of these conductors, the resistance of the tuyere maintains approximately constant small value regardless of its position in the metal. When the tip of the tuyere is positioned in the atmosphere of the furnace and in the layer of slag, the resistance of tuyere fluctuates; in the metal the fluctuations stop. Such changes in the tuyere resistance allow determining the moment when the tip gets in contact with the metal surface. When lowering down of the tuyere the minimum absolute value of resistance and stoppage of the fluctuations correspond just to the moment of its contacting the metal.

Automated control systems with the electrical resistance sensors have been developed by the Metallurgy Institute, city of Chelyabinsk, Russia, and successfully used for controlling the position of tuyeres relative to slag-metal interface in many open-hearth furnaces. By applying these sensors it was possible to keep the tips of the tuyeres near the metal surface with the 20–30 mm accuracy. This high accuracy is achieved due to the fact that during the blowing in the lower position of the tuyere its tip gets into contact with the metal. The use of similar systems in the shaft furnaces with melting a scrap in the liquid metal will significantly increase the bath blowing efficiency which will contribute to an increase in both the scrap melting rate and productivity.

References

1. Toulouevski Y, Zinurov I (2013) Innovation in electric arc furnaces: scientific basis for selection. Springer
2. Markov BL (1975) Blowing at a different position of the lances, Methods of the open-hearth bath blowing. Metallurgiya, Moscow, Russia, p 143

Chapter 6
High-Temperature Heating a Scrap in a Furnace Shaft

Abstract The productivity of a shaft furnace can be limited not only by the rate of scrap melting in the liquid metal but by the rate of scrap preheating as well. Heat transfer from gases to scrap in the shaft is carried out mainly by convection. The intensity of this process is mainly determined by the rate of gas flow in the shaft. An important feature of this flow is the vortices formed behind the transversely streamlined pieces of scrap. These vortices strongly turbulize the flow which dramatically increases the heat transfer coefficient from gases to scrap. An aerodynamic profile of such a flow is similar to the one that arises when the flow around a staggered tube bundle is transverse. This physical model is used to calculate the heating of equivalent scrap comprised of rods 25 mm in diameter staggered in the shaft. Cylindrical elements of an equivalent scrap 25 mm in diameter are "thin" bodies whose internal thermal resistance can be neglected. The calculations performed by using equations for heating of thin bodies to determine the heating time of the equivalent scrap in the Mexican furnace Quantum are given. The results of the calculations are close to the actual data. They show that it is the scrap heating that limits productivity of this furnace. The system of scrap preheating by the high-power recirculation burner devices is proposed to increase the productivity. The design and operation of the system are described. Due to recirculation the combustion temperature of natural gas with oxygen is reduced; the rates of gases passing through the scrap layer, the uniformity of their distribution along the shaft cross-section, and the intensity of heat transfer are sharply increased. All this contributes to achieving higher average mass temperatures of scrap preheating.

Keywords Scrap preheating rate in shaft · Calculation method · Scrap preheating with powerful recirculation burner devices

© The Author(s) 2017
Y.N. Toulouevski and I.Y. Zinurov, *Fuel Arc Furnace (FAF) for Effective Scrap Melting*, SpringerBriefs in Applied Sciences and Technology, DOI 10.1007/978-981-10-5885-1_6

6.1 Preliminary Considerations and Evaluation of Some Parameters

The productivity of a shaft furnace can be limited not only by the rate of scrap melting in the liquid metal but by the rate of scrap preheating in the shaft as well. Above it was shown that the scrap melting rates which are able to meet the highest requirements put forward by practice can be achieved by increasing the following parameters: the volume of the charging zone, the scrap preheating temperature, and the intensity of stirring of the bath. On the contrary, the possibilities of increasing the rate of scrap heating in the shaft are relatively small.

Heat transfer from gases to scrap is carried out mainly by convection. The intensity of this process is mainly determined by the rate of gas flow in the shaft filled with scrap. It is not possible to calculate some effective value of this rate and the corresponding heat transfer coefficient α at a chaotic location of various scrap pieces in the shaft. The necessary experimental data are also absent. Therefore, for analyzing the process it is necessary to substitute the equivalent model of scrap for the real scrap as in the case of determining the scrap melting rate in the liquid metal. An important feature of the flow of gases in the shaft is the vortices formed behind the transversely streamlined pieces of scrap. These vortices strongly turbulize the flow which despite its low velocity dramatically increases the heat transfer coefficient from gases to scrap. An aerodynamic profile of such a flow is similar to the one that arises when the flow around a staggered tube bundle is transverse, Fig. 6.1. In order to calculate the scrap heating process under conditions as close as possible to actual ones, the equivalent scrap should be located in the shaft so that its cylindrical elements would form a space transversely streamlined chess lattice similar to that is shown in Fig. 6.1.

In the Quantum furnace of 100 tons tapping weight, the area of cross-section of the shaft with dimensions of 4.6×2.6 m is about 12 m^2. The mass of a portion of the scrap placed on the fingers with $\rho = 0.65$ t/m^3 amounts to 27.5 tons, Chap. 4,

Fig. 6.1 Profile of fluid motion in the staggered tube bundle

Sect. 4.3.2.2. The equivalent scrap comprised of rods 0.025 m in diameter and length of 4.6 m will have the same mass if the rods are staggered on the fingers with gaps between the rods of 0.05 m both on the width of the shaft and its height.

The height of the layer of both actual and equivalent scrap in this case will be approximately 3.5 m and the area of a free section for passing gases through the layer of equivalent scrap will be 8 m². With an amount of gases passing through the shaft in similar furnaces of about 5 m³/s (s.t.p.) the rate of these gases will be 5/8 = 0.62 m/s. Such a low flow rate of off-gases must be accompanied by a drop in the heat transfer intensity from the gases to the scrap with an increase in the time of its heating to the desired temperature. Let us turn to the computational analysis of the corresponding dependences.

The heating of a fixed scrap layer by the flow of gases passing through it is a complex process. Calculations of this process in the framework of this book require significant simplifications. These simplifications are completely justified since, due to the lack of necessary experimental data, many of initial parameters of calculations can be estimated only approximately.

6.1.1 Calculation of Scrap Heating Time with off-Gases in the Quantum Shaft

According to the accepted classification of heated bodies cylindrical elements of an equivalent scrap 25 mm in diameter are "thin bodies". In the heating conditions corresponding to this concept the internal thermal resistance of thin bodies is neglected. It is assumed that the temperature of a thin body when it is heated is the same throughout the entire section. It can be assumed that the same is true for the heating of pieces of a real scrap of average granulometric composition.

The convection heating time of each thin cylindrical element of the equivalent scrap can be calculated by using the well-known equation:

$$\tau = 0.64 \frac{d \times \rho \times c_{st}}{2\alpha} lg\left(\frac{t_{gas} - t'_{st}}{t_{gas} - t''_{st}}\right) \tag{6.1}$$

τ is heating time, hr

$d = 0.025$ m

ρ is density of steel of 7900 kg/m³

c_{st} is average heat capacity of low-carbon steel in the temperature range of the heated body from the initial

t'_{st} to the final t''_{st}, J/(kg · K)

t_{gas} is local temperature of off-gases, °C

α is the coefficient of heat transfer from gases to scrap, W/(m² ·°C)

It is assumed that Eq. (6.1) can also be used to determine the heating time of the whole scrap portion located on the fingers of the shaft if to take as t_{gas} and t'' the average temperature of the gases passing through the scrap layer and the final mass average temperature of the scrap t_{st}^{av}., respectively. At a gas temperature at the entrance to the scrap layer of $t'_{gas} = 1650\ °C$ and that at the exit from the layer $t''_{gas} = 500\ °C$, the average temperature t_{gas}^{av} will be 1075 °C. The average mass temperature of scrap preheating $t_{st}^{av} = 400\ °C$. Within the range from 0 to 400 °C, $c_{st} = 0.540\ kJ/(kg\ ·°C)$.

In order to determine the heat transfer coefficient α a well-known equation for a transversely streamlined staggered tube bundle is used, Fig. 6.1. This equation represented by dimensionless numbers of Nusselt (Nu), Reynolds (Re) and Prandtl (Pr) has the form:

$$Nu = 0.40\,Re^{0.60} \times Pr^{0.36} \tag{6.2}$$

In the case under consideration $Nu = \alpha \times d/\lambda_{gas}$ and $Re = w_{gas} \times d/\nu_{gas}$.

Thermal conductivity of gases $\lambda_{gas} = 0.0896\ W/(m\ ·°C)$; $Pr_{gas} = 0.739$; kinematic viscosity of gases $\nu_{gas} = 170·10^{-6}\ m^2/s$; the actual gas velocity in a narrow section between the tubes $w_{gas} = 3.0\ m/s$. With such a determination of velocity the result of calculating α does not depend on the pitch between the tubes. This raises the degree of reliability of Eq. (6.2) when using it in calculating the time of scrap preheating. All the physical parameters were determined for the calculated values of the gas composition and temperature of $t_{gas}^{av} = 1075\ °C$. Using these parameters, with the help of the expressions for the dimensionless numbers Nu and Re and Eq. (6.2), we have determined the value of the coefficient $\alpha = 48.5\ W/(m^2\ °C)$. Since all the values entering into Eq. (6.1) are now known the time of heating can be determined: $\tau = 0.64 \frac{0.025 \times 7900 \times 0.540}{2 \times 48.5} lg\left(\frac{1075-0.0}{1075-400}\right) = 0.14$ h or 8.4 min. This time coincides with the interval between discharging of scrap portions weighing 27.5 tons from the shaft into the bath of a furnace operating in Mexico [1]. This fact confirms the sufficiently high accuracy of calculations. Thus, in this case, the heating of the scrap in the shaft is the bottleneck of the process since the heating time considerably exceeds the possible melting time of the scrap in the liquid metal, Chap. 4, Sect. 4.3.2.2. Equations (6.1) and (6.2) show that in order to reduce the scrap heating time τ it is necessary to increase both the rate and average temperature of the gas flow in the scrap layer. Such an opportunity is provided by the proposed patented system for scrap preheating in the furnace shaft.

6.2 Scrap Preheating System by High-Power Recirculation Burner Devices

The burners currently in use are unsuitable for high-temperature preheating a scrap in an EAF shaft. Air-gas burners of high power required generate too large amount of combustion products which impermissibly increase costs for their removal and purification. Oxy-gas burner flames have too high temperature, therefore, it is impossible to avoid undue oxidation, melting, and welding of scrap pieces. All this involves drop of the yield, suspension of scrap in the shaft, and underfiring of fuel, Chap. 2, Sect. 2.2.1. In order to avoid aforesaid shortcomings it is necessary to considerably reduce a temperature of combustion of natural gas with oxygen.

In the proposed system this problem is solved by means of recirculation of gases. Oxy-gas mixture generated by the burner devices is diluted inside them with those combustion products which have already passed through the layer of scrap, transferred heat energy to the scrap lowering, therefore, their own temperature. Such a recirculation of gases is being created by the oxygen injectors which are a part of the burner's devices. At the same time, intensive recirculation provides an increase in the gas flows rate in the furnace shaft which is necessary for the rapid preheating of scrap.

The water-cooled burner device is schematically shown in Fig. 6.2. The device comprises of oxygen chamber (1) distributing oxygen to several injectors. Each injector contains oxygen nozzle (2), mixing chamber (3), and diffuser (4). In chambers (3) oxygen mixes with the combustion products which after passing through the scrap layer are sucked into the injectors via openings (5). The injectors produce a positive pressure in chamber (6) into which multiple fine jets of natural gas are fed via nozzles (7). In the chamber (6) the natural gas, oxygen, and combustion products from the shaft are completely mixed.

The formed combustible mixture is blown under pressure via opening (8) into the furnace shaft where it is burned creating a flame with the low temperature required. The temperature of the flame decreases due to the presence in the combustible mixture of ballast in the form of the combustion products being cooled in

Fig. 6.2 Water-cooled burner device (designations are given in the text)

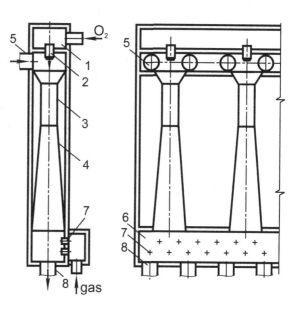

the shaft. The correct selections of technical parameters and quantity of oxygen injectors produce required excess of pressure in the chamber (6) as well as negative pressure in openings (5).

Installation of one of burner's devices (1) on the shaft of the Quantum furnace is shown in Fig. 6.3. The combustible mixture is introduced in lower scrap layers inside the shaft via pipes (2) close to fingers (3). The combustion products are sucked into the burner device via pipes (4). Thus, the combustion products before being sucked into the burner device pass through the scrap layer of a sufficient height which ensures required reduction in their temperature. Inflammation and complete combustion of the combustible mixture in the shaft is ensured due to the high temperature of the scrap, especially in its lower layer, and that of the off-gases. In order to eliminate backflash into the burner device the latter is equipped with flame arresters. A variant of a burner device with external mixing of gas with oxygen has been developed as well.

The flow of the gases in the upper part of the shaft is split, Fig. 6.3. The smaller part of it equal to the amount of combustion products produced by combusting the gas fuel with oxygen is removed from the shaft through the gas duct (5). The larger part of the flow is drawn into the burner devices by the oxygen injectors through the pipes (4) and then blown back into the shaft through the pipes (2). This part of the total flow of the gases circulates continually through the closed loop, i.e. the burner

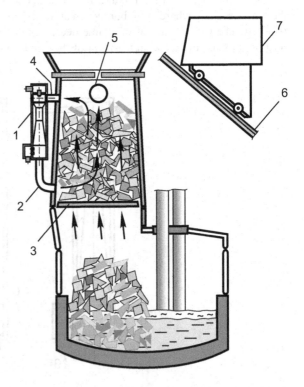

Fig. 6.3 Installation of burner device on a shaft at Quantum furnace (designations are given in the text)

devices—the layer of scrap—the burner devices. Due to the high power of the burner devices as well as to recirculation of the combustion products, the amount of gases passing through the scrap layer per unit time significantly exceeds the flow of off-gases leaving the furnace. The velocity of the gases passing through the layer of scrap, the uniformity of their distribution over the shaft cross-section, and the intensity of the heat transfer are much higher than those in the shaft furnaces which use only the off-gases for scrap preheating. All this contributes to reaching the higher average mass temperature of scrap preheating.

Reference

1. Apfel Jens, Mueller A et al (2016) EAF Quantum—Results of 2015, EEC 2016, Proceedings, Venice, Italy

Chapter 7
Fuel Arc Furnace—FAF

Abstract Selection of the Quantum constructive scheme as a basis for a fuel arc furnace—FAF is justified. Main calculated parameters and performances of the FAF of 100 tons tapping weight equipped with the proposed systems for high-temperature scrap preheating and oxygen bath blowing at the slag-metal interface are given. The methods described earlier are used for the calculations. The heating time of scrap in the shaft to an average mass temperature of 800 °C and conditions required for such heating are defined. The powers of the transformer and burner devices, electric energy consumption and natural gas flow rate have been determined as well. The hourly productivity of the FAF increases by a factor of 1.4 and electrical energy consumption reduces by about 2 times in comparison with the EAF with the same tapping weight. This effect is reached with the 30% reduced power of the transformer and the increased natural gas flow rate by 5.5 m^3/t. These and other advantages of the FAF make it possible to expect that in the near future the fuel arc furnaces will be able not only to compete successfully with the modern EAFs but also replace them everywhere.

Keywords Concept of fuel arc furnace (FAF) · Calculated performances of 100-t Quantum-type FAF · Technical, economical and ecological advantages of FAF

7.1 Concept of the Fuel Arc Furnace

The fundamental features of the concept of the fuel arc furnace FAF are:

- High-temperature scrap heating in the shaft by powerful recirculation burner devices. Such heating allows to dramatically reduce electrical energy consumption and significantly increase productivity of the furnace without increasing the transformer power
- Intensive recirculation of gases in the shaft in a closed loop: scrap layer—burner device—scrap layer. The recirculation reduces the combustion temperature of an oxy-gas mixture in the scrap layer and shortens the time of scrap heating to high temperatures. This is achieved due to a sharp increase in both the velocity of the

Y.N. Toulouevski and I.Y. Zinurov, *Fuel Arc Furnace (FAF) for Effective Scrap Melting*, SpringerBriefs in Applied Sciences and Technology, DOI 10.1007/978-981-10-5885-1_7

gases in the shaft and the heat transfer coefficient α from gases to scrap without an increase in the volume of gases requiring purification. The promoters of recirculation are the injectors which are part of the burner devices and use the energy of oxygen as well as nitrogen or compressed air

- Increase in the rate of scrap melting in a liquid metal due to an increase in the volume of the scrap charging zone and also intensive stirring of the metal between this zone and the electric arcs zone. This is achieved by using oxygen lances immersed into the melt to the slag-metal interface.

Deep replacement of electrical energy with fuel energy and convergence of the power of the burner devices with the power of the transformer explain the choice of the name for the new steelmaking unit: "Fuel Arc Furnace FAF".

7.1.1   Selection of the Quantum Constructive Scheme as a Base for FAF

As already mentioned, at present, very favorable conditions have been created for the implementation of the FAF concept due not only to a sharp increase in the available resources of relatively cheap shale gas but also to the creation of several types of shaft furnaces with the melting of scrap in a liquid metal bath. The designs of these furnaces are mastered in practice and can be used as a base for the FAF. These include furnaces with fingers as Quantum, COSS furnaces with pushers, and twin-shell shaft furnaces. Let us examine their advantages and disadvantages in terms of using those as a basic scheme for the FAF.

The advantage of the COSS furnaces is the charging of the bath in small consecutive scrap portions. This charging is essentially the same as the continuous conveyer charging on the Consteel furnaces with all its advantages. In addition, with each portion, the pusher discharges from the shaft the lowest layers of the scrap heated to the highest temperature. This increases the average temperature of the scrap charged into the bath.

However, operating experience has shown that in the COSS furnaces the problem of reliable operation of pushers was not completely solved even at average mass temperatures of scrap preheating not exceeding 450 °C. These problems increase significantly at higher temperatures. Hydrocylinders and pushers are water-cooled. In addition, water is used for spraying the bottom of the shaft which facilitates the pushers operation. The presence of water in the discharging zone of the scrap from the shaft creates potentially dangerous situations, especially in the case of leakages of water from the cooled elements. Perhaps, this is why in ECOARC furnaces the pushers are cooled by compressed air which excludes the possibility of high-temperature scrap preheating. Insufficiently reliable operation of the pusher does not allow at present to recommend the use of a design of the COSS furnaces for the FAF despite their aforesaid advantages.

Twin-shell shaft furnaces create very favorable conditions for high-temperature scrap preheating; however, they require considerable additional costs for their construction and maintenance. Therefore, preference should be given to shaft finger furnaces of the Quantum-type although the charging of scrap into the bath with several large portions is not optimal. This disadvantage is compensated by reliable operation of the furnaces with fingers proved by long-term practice. The requirements of the FAF are most fully met the design of the Quantum furnace thanks to the most advanced finger system and most promising scrap charging system as well.

7.1.2 Calculations of Main Parameters and Performances of the FAF

7.1.2.1 Data on Parameters and Operating Conditions of the furnace Required for Calculations

A Quantum furnace of 100 tons tapping weight and with the average charging zone volume in the course of melting of 7.5 m^3 was selected for the calculations. The Quantum furnace operating in Mexico has such parameters. This allows us to use in calculations the data obtained earlier in the analysis of the operation of this furnace, Chap. 4, Sect. 4.3.2.2 and Chap. 6, Sect. 6.1.1. It is assumed that at the FAF furnace under consideration the scrap is preheated up to an average mass temperature of 800 °C. To heat the scrap, natural gas is used together with off-gases.

In order to shorten the time of scrap preheating to a temperature of 800 °C the distribution of gas flows and temperature along the height of the scrap layer in the shaft is changed. The combustible mixture of gases generated by burner devices is introduced not only in the lower but also in the upper scrap layers. All this makes it possible to raise the temperature of the gases at the outlet of the shaft to about 1150 °C and the average temperature of the gases passing through the scrap layer to 1400 °C. Such an increase in gas temperatures solves two problems: increases the rate of scrap preheating and ensures complete decomposition of dioxins at the outlet from the shaft. The necessary increase in the flow rate of natural gas not only provides an additional heating of the scrap, but also is completely compensated for by the absence of a special high-temperature chamber for the re-heating of gases and for decomposing the dioxins contained in them. At low temperatures of gases leaving the shaft such chambers are necessary and widely used. The gas flow rate in these cumbers is 4–5 m^3/t.

The combustion of natural gas in the shaft, the use of not only oxygen but also nitrogen in the injectors, and an increase in the average temperature of gases to 1400 °C are the factors which sharply increase the rate of gases in the shaft and shorten the time of scrap preheating.

7.1.2.2 Calculation of Scrap Preheating Time

The same calculation method which has already been applied to the Quantum furnace is used. All physical parameters of the mixture of natural gas combustion products with off-gases are determined at a temperature t_{gas} of 1400 °C. By using these parameters the dimensionless numbers of Re = 987×10^3 and Nu = 0.153 were determined. Substituting these numbers into Eq. (6.2) we find α = 148 W/($m^2 \cdot °C$). It should be considered that at the high-temperature preheating of scrap the coefficient α increases by about 10% due to the radiation of heavily dusted gases. Taking into account the radiation the total coefficient α will be $148 \times 1.1 = 163$ W/($m^2 \cdot °C$).

The time of heating the scrap to 800 °C τ_{800} is determined considering the oxidation of 1.5% of iron to Fe_3O_4. The amount of heat released during this process can heat the scrap from 700 to 800 °C. Therefore, to determine the time of heating the scrap to a temperature of 800 °C, in Eq. (6.1), Chap. 6, a temperature of 700 °C but not 800 °C can be assumed as the final heating temperature. The average heat capacity of steel in the temperature range from 0 to 800 °C c_{st} = 0.695 kJ/(kg °C). Let us find τ_{800} according to Eq. (6.1): $\tau_{800} = \frac{0.025 \times 7900 \times 0.695}{2 \times 163} lg\left(\frac{1400-0}{1400-700}\right) = 0.081$ hr or $\tau_{800} \cong 5$ minutes. The total heating time of the four scrap portions will be 20 minutes.

7.1.2.3 Required Transformer Power and Electrical Energy Consumption

Scrap melting period. During this period the transformer should provide a rational melting rate equal to the rate of scrap heating in the shaft since calculations show that it is the heating of the scrap in the shaft that is, in this case, the bottleneck of the entire process of the heat. On the considered FAF, the rational melting time of 27.5 tons of scrap is 5 min. In order to melt down 1 ton of scrap and heat the melt to a temperature of 1580 °C it is necessary to input to the furnace bath 379 kWh/t, Chap. 4, Sect. 4.3.2.2. Exothermic reactions of oxidation of iron and its alloys contribute 79 kWh/t, preheated to 800 °C scrap contributes 145 kWh/t and coke—34 kWh/t. Electrical energy has to contribute for 5 min (0.083 h) 379—(79 + 145 + 34) = 121 kWh/t or 121 × 27.5 = 3327 kWh. With the electrical energy efficiency coefficient η_{el} = 0.84 and $cos\varphi$ = 0.79 a requited power of the transformer during the melting period amounts to 3327/ $(0.083 \times 0.84 \times 0.79 \times 10^3)$ = 60.4 MVA.

Liquid metal heating period to a tapping temperature. The duration of the period is 4 min (0.067 h); the mass of the liquid metal considering the hot heel mass is 160 tons; the metal is heated from 1580 to 1640 °C. The amount of heat which is introduced into the bath for 0.067 h is $(E_{1640}-E_{1580}) \times 160$; (394 $-379) \times 160 = 2400$ kWh. A required power of the transformer is 2400/ $(0.067 \times 0.84 \times 0.79 \times 10^3)$ = 60.2 MVA. Taking into account additional unaccounted energy losses a transformer with a power of 70 MVA is required for the FAF under consideration.

Electrical energy consumption. The value of consumption is determined by using Curve 2 in Fig. 2.8, Chap. 2, Sect. 2.3.2.1. It amounts to approximately 200 kWh/t.

7.1.2.4 Power of Burner Devices and Natural Gas Flow Rate

The amount of heat Q that burner devices must introduce into the shaft for 5 min (0.083 h) to heat 27.5 tons of scrap to 800 °C is determined by the expression: $Q = (E_{800}-E_{400}-E_{ch}) \times 27.5/\eta_{br}$, kWh, where E_{800} = 145 kWh/t is the enthalpy of scrap at 800 °C; E_{400} = 58 kWh/t is the enthalpy of scrap at 400 °C (the heating to 400 °C is produced by off-gases); E_{ch} = 28 kWh/t is the amount of heat released when oxidizing 1.5% of iron to Fe_3O_4; 27.5 tons is the mass of a heated scrap portion; η_{br} = 0.6 is the efficiency coefficient of natural gas in the shaft. This value decreases from 0.7 to 0.6 due to an increase in the temperature of the gases at the outlet from the shaft to 1150 °C for the decomposition of dioxins. Q = (145−58 −28) × 27.5/0.6 = 2704 kWh. The power of burner devices is P_{br} = 2704/0.083 $\cong$ 32.6 MW.

Natural gas flow rate. With the gas calorific value of 10.3 kWh/m^3 the gas flow rate is 2704 × 4/(10.3×100) = 10.5 m^3/t.

7.1.2.5 Tap-to-Tap Times and Hourly Productivity

In the Quantum furnace the melting time of each of the four scrap portions is 7 min. In the FAF this time is taken equal to the time of heating a portion of the scrap and amounts to 5 min. The melting time of the entire scrap in the FAF is reduced by 8 min. With the same duration of other periods of the heat the tap-to-tap time is shortened from 36 min in the Quantum to 28 min in the FAF.

7.2 Advantages of Fuel Arc Furnaces FAF of Quantum-Type

Performances of a modern EAF, Quantum furnace, and FAF of Quantum-type are given in Table 7.1. These data indicate the undeniable technical, economic and environmental benefits of the FAF. With the same tapping weight the hourly productivity of the FAF is higher than that of EAF by a factor of 1.4, and the electrical energy consumption is about twice lower than that in the EAF, Table 7.1. This effect is reached with the reduced power of the transformer by 30% and the increased natural gas flow rate by 5.5 m^3/t.

With current prices in the US for gas of $ 0.12 per 1 m^3 and for electricity of $ 0.06 per 1 kWh, the reduction in total cost of these energy carriers in the FAF compared to that the EAF is about $ 8.5/t, including the cost of oxygen. In addition,

Table 7.1 Performances of EAF, Quantum, and FAF

Performances	EAF	Quantum	FAF
Tapping weight, t	100	100	100
TTT time, min	40	36	28
Productivity, t/h	149	167	214
Electrical energy consumption, kWh/t	375	280	200
Transformer power, MVA	100	80	70
Total power of burner devices, MW	–	–	33
Scrap preheating temperature, °C	–	400	800
Natural gas flow rate for scrap, m^3/t	5.0	4.4	10.5

absolute savings of gas in cubic meters also take place in the system FAF—thermal power station (TPS) supplying electricity to the furnace. In TPS 2.9 m^3 of gas is saved per each 1 m^3 of gas consumed in the FAF for scrap preheating. Accordingly, CO_2 emissions into the atmosphere in the FAF—TPS system are reduced by about half [1].

In conclusion, it should be emphasized once again that the creation of the FAF in the coming years became possible due to the two fundamental innovations: scrap charging into a liquid metal bath and scrap preheating in shaft furnaces. These innovations are associated with the names of J. Vallomy and G. Fuchs as well as with the companies Tenova, Fuchs Technology, and Primetals. Advantages of the FAF allow to expect that in the near future the fuel arc furnaces will be able not only to successfully compete with the modern EAFs but also replace them everywhere.

Reference

1. Toulouevski YN, Zinurov IY (2015) Electric arc furnace with flat bath. Achievements and Prospects, Springer

Index

© The Author(s) 2017
Y.N. Toulouevski and I.Y. Zinurov, *Fuel Arc Furnace (FAF) for Effective
Scrap Melting*, SpringerBriefs in Applied Sciences and Technology,
DOI 10.1007/978-981-10-5885-1

Printed in the United States
By Bookmasters